Problem-Solving Guide with Solutions
to accompany

COLLEGE PHYSICS

Volume II

Timothy A. French
DePaul University

W. H. FREEMAN AND COMPANY
New York

© 2014 by W. H. Freeman and Company
All rights reserved.
Printed in the United States of America

ISBN-13: 978-1-4641-0138-0
ISBN-10: 1-4641-0138-8

First printing

W. H. Freeman and Company
41 Madison Avenue
New York, NY 10010
RG21 6XS England

www.whfreeman.com/physics

Contents

Volume II

16 Electrostatics I: Electric Charge, Forces, and Fields   1
17 Electrostatics II: Electric Potential Energy and Electric Potential   16
18 Electric Charges in Motion   27
19 Magnetism   33
20 Electromagnetic Induction   42
21 Alternating-Current Circuits   46
22 Electromagnetic Waves   56
23 Wave Properties of Light   61
24 Geometrical Optics   76
25 Relativity   93
26 Quantum Physics and Atomic Structure   104
27 Nuclear Physics   117
28 Particle Physics   127

About the Author

DR. TIMOTHY FRENCH

Dr. Timothy French is currently a Visiting Assistant Professor in the Department of Chemistry at DePaul University in Chicago, IL. Dr. French holds a BS in chemistry from Rensselaer Polytechnic Institute and a MS and PhD in chemistry from Yale University. His undergraduate research looked at how to assess problem-solving ability in introductory physics students, while his doctoral research was on terahertz spectroscopy. Prior to arriving at DePaul, he was a Preceptor in Chemistry and Chemical Biology for four years at Harvard University where he taught an introductory physics sequence with biological and medical applications, experimental physical chemistry, and introductory physical chemistry, as well as the introductory physics class at the Harvard Summer School. He was awarded the Harold T. White Prize for Excellence in Teaching from the Harvard Physics Department in 2009.

A Note from the Author

The Problem-Solving Guide with Solutions is more than a book of answers to physics problems. It's designed to help you learn the way physicists approach and solve problems.

Many students, when approaching a physics problem, will half-read the question, pick out some of the variables used in the problem statement, and then start blindly hunting for an equation that shares the same letters, regardless of its utility.

Physicists, on the other hand, approach problems very differently. They'll fully read the question, underlining important information as they go. Next they will draw their own picture, even if one is provided, in order to quickly and easily organize and process the information provided. Then they will consider the underlying concept at play ("Is this a conservation of momentum question? Is Newton's second law best used here?") before determining which equation will be the most helpful. The equations are all manipulated algebraically before putting in numbers and arriving at a numerical answer.

Students usually move on to the next problem at this point without performing the crucial final step—checking the answer for reasonability. Physicists make sure the answer has the correct units, check some limiting cases ("What happens if the angle goes to 0 degrees? 90 degrees?"), and perform order of magnitude estimations all in order to gain confidence in their final answer.

Each solution in this book is set up to mirror this process with three distinct parts: **Set Up, Solve,** and **Reflect**. The **Set Up** portion contains all of the logic behind starting the problem, from a rehash of the important information in the problem statement to the conceptual underpinnings necessary in understanding the solution. The Solve portion contains the algebraic steps used to arrive at the numerical solution; note that this step comes *after* the Set Up step. Finally, the Reflect step provides a "sanity check" for the answer—"Is this what we expected? Does it make sense with respect to our observations of and interactions with everyday life? Does this answer seem reasonable?"

Very often instructors expect students to become expert problem solvers on their own without instruction. By explicitly showing the steps physicists use when solving problems, I hope you will be able to mirror and internalize these steps and make your foray into physics a more enjoyable and successful experience.

—Tim

Get Help with Premium Multimedia Resources

One of the benefits technology brings us in education is the ability to visualize concepts, gain problem support, and test our skills outside of the traditional pen-and-paper classroom method. With that in mind, W. H. Freeman has developed a series of media assets geared at reinforcing conceptual understanding and building problem-solving skills.

P'Casts are videos that emulate the face-to-face experience of watching an instructor work a problem. Using a virtual whiteboard, the P'Casts' tutors demonstrate the steps involved in solving key worked examples, while explaining concepts along the way. The worked examples were chosen with the input of physics students and instructors across the country. P'Casts can be viewed online or downloaded to portable media devices.

Interactive Exercises are active learning, problem-solving activities. Each Interactive Exercise consists of a parent problem accompanied by a Socratic-dialog "help" sequence designed to encourage critical thinking as users do a guided conceptual analysis before attempting the mathematics. Immediate feedback for both correct and incorrect responses is provided through each problem-solving step.

Picture Its help bring static figures from the text to life. By manipulating variables within each animated figure students visualize a variety of physics concepts. Approximately 50 activities are available.

These Premium Multimedia Resources are available to you in a few places:
- If you're using the printed text, you can purchase the Premium Media Resources for a small fee via the Book Companion Website (BCS) at www.whfreeman.com/collegephysics1e.
- If you've purchased the W. H. Freeman media-enhanced eBook, these resources are embedded directly into the chapters of the text and available on the BCS.
- If you're using an online homework system, such as WebAssign, these resources are integrated into your individual assignments and are available on the BCS.

And while all three of those places correlate the Premium Media Resources by chapter, we've gone one step further in the Problem-Solving Guide with Solutions and correlated each media resource by problem. Look for each relevant item called out after its corresponding problem:

> **Get Help:** Picture It - Adding and Subtracting Vectors

To get started using these exciting resources, log on to www.whfreeman.com/collegephysics1e!

Chapter 16
Electrostatics I: Electric Charge, Forces, and Fields

Conceptual Questions

16.5   The mass decreases because electrons are removed from the object to make it positively charged.

   Get Help: P'Cast 16.1 – Electrons in a Raindrop

16.9   When the plastic comb is run through your hair, electrons are transferred from your hair to it, so the comb acquires a net negative charge. This charge polarizes the molecules in the paper. The ends of the molecules closest to the comb have a slight positive charge, and the ends of the molecules far from the comb have an equal amount of negative charge. Because the positive ends of the molecules are closer to the comb than the negative ends, the net force on the molecules, and hence the paper, is attractive. When the comb touches the paper, some electrons are transferred from the comb to the paper, giving the paper a net negative charge. Because the paper now has a net negative charge, the comb and paper are repelled from each other.

16.13   Part a) If the charges creating the field move, the fact that the field propagates at the speed of light also allows us to understand the changes in the field, and hence force on the other charged objects. With the electric field, we see that charged particles take some time to experience the effects of other charges moving near them.

   Part b) In electrostatics the field is just a computational device, and using it is merely a matter of convenience. However, in electrodynamics the field is necessary for energy and momentum to be conserved, so it is more than just convenience that leads us to the electric field.

   Get Help: Picture It – Electric Field
   P'Cast 16.2 – Calculating Electric Force
   P'Cast 16.6 – Where Is the Electric Field Zero?

16.17   Only charges inside the Gaussian surface contribute a nonzero flux, so the electric field of charges outside the surface can be ignored.

Multiple-Choice Questions

16.21   **C** (F). The forces must have the same magnitude due to Newton's third law.

16.23   **A** (positively). Charge can move easily in a conductor, so the mobile electrons will be attracted to the positively charged rod, leaving the other end positively charged.

16.27   **B** ($-Q$). The electric field inside the conducting shell must be zero. The charge enclosed by a Gaussian sphere of radius r, where $R_1 < r < R_2$, must be zero. Therefore, the charge on the inner surface of the shell is $-Q$.

   Get Help: P'Cast 16.18 – Gauss' Law and Surface Charge

Estimation/Numerical Analysis

16.31

$$F = \frac{k|q_1||q_2|}{r^2}$$

$$q \sim \sqrt{\frac{(10 \text{ N})(10^{-1} \text{ m})^2}{\left(10^{10} \frac{\text{N} \cdot \text{m}^2}{\text{C}^2}\right)}} = 10^{-5} \text{ C}$$

Problems

16.37

SET UP

We are asked to calculate the number of coulombs of negative charge in 0.500 kg of water. To answer this, we need to know the total number of electrons in the sample. The molar mass of water is 18.0 g/mol, and 1 mol of water contains 6.022×10^{23} molecules of water. Each molecule of water is made up of two hydrogen atoms and one oxygen atom. Hydrogen has one electron and oxygen has eight, which means each water molecule contains 10 electrons. The charge on one electron is -1.602×10^{19} C.

SOLVE

$$0.500 \text{ kg} \times \frac{1000 \text{ g}}{1 \text{ kg}} \times \frac{1 \text{ mol}}{18.0 \text{ g}} \times \frac{6.022 \times 10^{23} \text{ molecules}}{1 \text{ mol}} \times \frac{10 \text{ electrons}}{1 \text{ molecule}} \times \frac{-1.602 \times 10^{-19} \text{ C}}{1 \text{ electron}}$$

$$= \boxed{-2.68 \times 10^7 \text{ C}}$$

REFLECT

The net charge of 0.5 kg of water is zero since there are also 10 protons per molecule.

Get Help: P'Cast 16.1 – Electrons in a Raindrop

16.43

SET UP

A person walking across a carpet accumulates -50 nC of charge in each step. Multiplying this value by 25 will tell us the amount of charge she builds up in 25 steps. Dividing the total charge on the person by the charge on a single electron (-1.602×10^{-19} C) yields the total number of excess electrons present due to static buildup. Finally, from these same values, we can calculate the number of steps required to accumulate a total of 10^{12} electrons.

SOLVE

Part a)

$$25 \text{ steps} \times \frac{-50 \text{ nC}}{1 \text{ step}} = \boxed{-1250 \text{ nC} = -1.25 \text{ }\mu\text{C} = -1 \times 10^{-6} \text{ C}}$$

Part b)

$$-1.25 \times 10^{-6} \text{ C} \times \frac{1 \text{ electron}}{-1.602 \times 10^{-19} \text{ C}} = \boxed{8 \times 10^{12} \text{ electrons}}$$

Part c)

$$10^{12} \text{ electrons} \times \frac{-1.602 \times 10^{-19} \text{ C}}{1 \text{ electron}} \times \frac{1 \text{ step}}{-50 \times 10^{-9} \text{ C}} = 3.2 \text{ steps}$$

A worker $\boxed{\text{should not take more than 3 steps}}$ before touching the components.

REFLECT

People building electronics or working with precise instrumentation will "ground" themselves often by touching a metal conduit or wearing an antistatic wrist strap to rid themselves of excess charge buildup.

16.45

SET UP

We can use Coulomb's law to calculate the distance between two electrons such that the electrostatic force exerted by each on the other was equal in magnitude to the force of gravity on an electron. The mass of an electron is $m_{electron} = 9.11 \times 10^{-31}$ kg; the magnitude of an electron's charge is $e = 1.602 \times 10^{-19}$ C.

SOLVE

$$F_{q_1 \text{ on } q_2} = \frac{k|q_1||q_2|}{r^2} = w_{electron} = m_{electron}g$$

$$\frac{k(e)(e)}{r^2} = m_{electron}g$$

$$r = e\sqrt{\frac{k}{m_{electron}g}} = (1.602 \times 10^{-19} \text{ C})\sqrt{\frac{\left(8.99 \times 10^9 \frac{\text{N} \cdot \text{m}^2}{\text{C}^2}\right)}{(9.11 \times 10^{-31} \text{ kg})\left(9.80 \frac{\text{m}}{\text{s}^2}\right)}} = \boxed{5.08 \text{ m}}$$

REFLECT

The electric force between two electrons is repulsive, while the gravitational force between an electron and Earth is attractive.

16.49

SET UP

Two charges, $q_1 = +5.00$ nC and $q_2 = -7.00$ nC, are located along the x-axis at $x = 0$ and $x = 5.00$ m, respectively. A third charge, $q = +2.00$ nC, is placed at a position x, where the net force acting on the charge C is zero. To find this position, we can set the magnitude of the electric force of q_1 on the charge equal to the magnitude of the electric force of q_2 on the charge and solve for x.

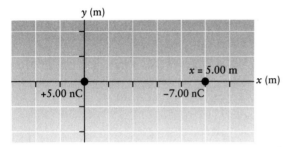

Figure 16-1 Problem 49

SOLVE

$$F_{q_1 \text{ on } q} = \frac{k|q_1||q|}{r_{q_1}^2} \quad \text{and} \quad F_{q_2 \text{ on } q} = \frac{k|q_2||q|}{r_{q_2}^2}$$

$$F_{q_1 \text{ on } q} = F_{q_2 \text{ on } q}$$

$$\left|\frac{k|q_1||q|}{x_{q_1}^2}\right| = \left|\frac{k|q_2||q|}{x_{q_2}^2}\right|$$

$$\frac{|q_1|}{x^2} = \frac{|q_2|}{(x-5.00)^2} \quad \text{(assume distances are in m, charges in nC)}$$

$$(|q_1| - |q_2|)x^2 - 10.0|q_1|x + 25.0|q_1| = 0$$

$$x = \frac{10.0|q_1| \pm \sqrt{(10.0|q_1|)^2 - 4(|q_1| - |q_2|)(25.0|q_1|)}}{2(|q_1| - |q_2|)}$$

$$x = \frac{50.0 \pm \sqrt{2500 - 4(-2.00)(125)}}{2(-2.00)} = \frac{50.0 \pm \sqrt{3500}}{-4.00}$$

Taking the positive root:

$$x = \frac{50.0 + \sqrt{3500}}{-4.00} = \boxed{-27.3 \text{ m}}$$

REFLECT

The $(x - 5.00)^2$ in the denominator of $F_{q_2 \text{ on } q}$ represents the fact that q_2 is located 5.00 m to the right of the origin. It makes sense that q must be placed to the left of q_1 since the magnitude of q_2 is larger than the magnitude of q_1 and $\vec{F}_{q_1 \text{ on } q}$ is repulsive, while the $\vec{F}_{q_2 \text{ on } q}$ is attractive.

16.51

SET UP

Three positive charges ($q_A = 3.00 \times 10^{-6}$ C, $q_B = 6.00 \times 10^{-6}$ C, $q_C = 2.00 \times 10^{-6}$ C) are placed in a coordinate system (see figure). We can use Coulomb's law, $F_{q_1 \text{ on } q_2} = \dfrac{k|q_1||q_2|}{r^2}$, to calculate the net force acting on each charge due to the two others. Since all of the charges are positive, all of the forces will be repulsive.

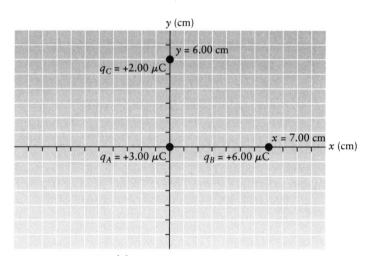

Figure 16-2 Problem 51

SOLVE

Charge A:

x component of net force on charge A:

$$\sum F_{\text{net on } A,x} = F_{B \text{ on } A,x} + F_{C \text{ on } A,x} = F_{B \text{ on } A,x} + 0 = \frac{k|q_B||q_A|}{r_{AB}^2}$$

$$= \frac{\left(8.99 \times 10^9 \dfrac{\text{N} \cdot \text{m}^2}{\text{C}^2}\right)|6.00 \times 10^{-6} \text{ C}||3.00 \times 10^{-6} \text{ C}|}{(7.00 \times 10^{-2} \text{ m})^2}$$

$$= 33.02 \text{ N} = 33.0 \text{ N (rounded to 3 significant figures)}$$

Because charges A and B have the same sign, the force on charge A is repulsive and acts in the negative x direction. Thus, the net force on charge A in the x direction is -33.0 N.

y component of net force on charge A:

$$\sum F_{\text{net on }A,y} = F_{B \text{ on }A,y} + F_{C \text{ on }A,y} = 0 + F_{C \text{ on }A,y} = \frac{k|q_C||q_A|}{r_{AC}^2}$$

$$= \frac{\left(8.99 \times 10^9 \frac{\text{N} \cdot \text{m}^2}{\text{C}^2}\right)|2.00 \times 10^{-6} \text{ C}||3.00 \times 10^{-6} \text{ C}|}{(6.00 \times 10^{-2} \text{ m})^2}$$

$$= 14.98 \text{ N} = 15.0 \text{ N (rounded to 3 significant figures)}$$

Because charges A and C have the same sign, the force on charge A is repulsive and acts in the negative y direction. Thus, the net force on charge A in the y direction is -15.0 N.

Magnitude of net force on charge A:

$$F_{\text{net on }A} = \sqrt{(F_{\text{net on }A,x})^2 + (F_{\text{net on }A,y})^2}$$

$$= \sqrt{(-33.02 \text{ N})^2 + (-14.98 \text{ N})^2} = \boxed{36.3 \text{ N}}$$

Direction of net force on charge A:

$$\theta_A = \tan^{-1}\left(\frac{-14.98 \text{ N}}{-33.02 \text{ N}}\right) = 24.4°$$

Since the force vector points in the third quadrant, the direction of the net force on charge A is $180° + 24.4° = \boxed{204.4°}$.

Charge B:

x component of net force on charge B:

$$\sum F_{\text{net on }B,x} = F_{A \text{ on }B,x} + F_{C \text{ on }B,x} = \frac{k|q_A||q_B|}{r_{AB}^2} + \frac{k|q_C||q_B|}{r_{CB}^2}\cos\theta_{CB}$$

$$= \frac{k|q_A||q_B|}{r_{AB}^2} + \frac{k|q_C||q_B|}{(r_{AB}^2 + r_{AC}^2)}\left(\frac{r_{AB}}{\sqrt{r_{AB}^2 + r_{AC}^2}}\right)$$

$$= \frac{k|q_A||q_B|}{r_{AB}^2} + \frac{k|q_C||q_B|r_{AB}}{(r_{AB}^2 + r_{AC}^2)^{3/2}}$$

$$= \frac{\left(8.99 \times 10^9 \frac{\text{N} \cdot \text{m}^2}{\text{C}^2}\right)|3.00 \times 10^{-6}\,\text{C}||6.00 \times 10^{-6}\,\text{C}|}{(7.00 \times 10^{-2}\,\text{m})^2}$$

$$+ \frac{\left(8.99 \times 10^9 \frac{\text{N} \cdot \text{m}^2}{\text{C}^2}\right)|2.00 \times 10^{-6}\,\text{C}||6.00 \times 10^{-6}\,\text{C}|(7.00 \times 10^{-2}\,\text{m})}{[(7.00 \times 10^{-2}\,\text{m})^2 + (6.00 \times 10^{-2}\,\text{m})^2]^{3/2}}$$

$$= 42.66\,\text{N} = 42.7\,\text{N}\ (\text{rounded to 3 significant figures})$$

y component of net force on charge B:

$$\sum F_{\text{net on }B,y} = F_{A\text{ on }B,y} + F_{C\text{ on }B,y} = 0 + \frac{k|q_C||q_B|}{r_{BC}^2}\sin\theta_{CB}$$

$$= \frac{k|q_C||q_B|}{r_{AB}^2 + r_{AC}^2}\left(\frac{r_{AC}}{\sqrt{r_{AB}^2 + r_{AC}^2}}\right)$$

$$= \frac{k|q_C||q_B|r_{AC}}{(r_{AB}^2 + r_{AC}^2)^{3/2}}$$

$$= \frac{\left(8.99 \times 10^9 \frac{\text{N} \cdot \text{m}^2}{\text{C}^2}\right)|2.00 \times 10^{-6}\,\text{C}||6.00 \times 10^{-6}\,\text{C}|(6.00 \times 10^{-2}\,\text{m})}{[(7.00 \times 10^{-2}\,\text{m})^2 + (6.00 \times 10^{-2}\,\text{m})^2]^{3/2}}$$

$$= 8.260\,\text{N} = 8.26\,\text{N}\ (\text{rounded to 3 significant figures})$$

Because charges C and B have the same sign, the force on charge B is repulsive and acts in the negative y direction. Thus, the net force on charge B in the y direction is -8.26 N.

Magnitude of net force on charge B:

$$F_{\text{net on }B} = \sqrt{(F_{\text{net on }B,x})^2 + (F_{\text{net on }B,y})^2}$$

$$F_{\text{net on }B} = \sqrt{(42.7\,\text{N})^2 + (-8.26\,\text{N})^2} = \boxed{43.5\,\text{N}}$$

Direction of net force on charge B:

$$\theta_B = \tan^{-1}\left(\frac{-8.260\,\text{N}}{42.66\,\text{N}}\right) = -11.0°$$

Since the force vector points in the fourth quadrant, the direction of the net force on charge B is $360° - 11.0° = \boxed{349.0°}$.

8 Chapter 16 Electrostatics I: Electric Charge, Forces, and Fields

Charge C:

x component of net force on charge C:

$$\sum F_{\text{net on C},x} = F_{A \text{ on C},x} + F_{B \text{ on C},x} = 0 + \frac{k|q_B||q_C|}{r_{BC}^2} \cos \theta_{BC}$$

$$= \frac{k|q_B||q_C|}{r_{AB}^2 + r_{AC}^2} \left(\frac{r_{AB}}{\sqrt{r_{AB}^2 + r_{AC}^2}} \right)$$

$$= \frac{k|q_B||q_C|r_{AB}}{(r_{AB}^2 + r_{AC}^2)^{3/2}}$$

$$= \frac{\left(8.99 \times 10^9 \frac{\text{N} \cdot \text{m}^2}{\text{C}^2}\right)|6.00 \times 10^{-6} \text{ C}||2.00 \times 10^{-6} \text{ C}|(7.00 \times 10^{-2} \text{ m})}{[(7.00 \times 10^{-2} \text{ m})^2 + (6.00 \times 10^{-2} \text{ m})^2]^{3/2}}$$

$$= 9.636 \text{ N} = 9.64 \text{ N (rounded to 3 significant figures)}$$

Because charges C and B have the same sign, the force on charge C is repulsive and acts in the negative x direction. Thus, the net force on charge C in the x direction is -9.64 N.

y component of net force on charge C:

$$\sum F_{\text{net on C},y} = F_{A \text{ on C},y} + F_{B \text{ on C},y} = \frac{k|q_A||q_C|}{r_{AC}^2} + \frac{k|q_B||q_C|}{r_{BC}^2} \sin \theta_{BC}$$

$$= \frac{k|q_A||q_C|}{r_{AC}^2} + \frac{k|q_B||q_C|}{r_{AB}^2 + r_{AC}^2} \left(\frac{r_{AC}}{\sqrt{r_{AB}^2 + r_{AC}^2}} \right)$$

$$= \frac{k|q_A||q_C|}{r_{AC}^2} + \frac{k|q_B||q_C|r_{AC}}{(r_{AB}^2 + r_{AC}^2)^{3/2}}$$

$$= \frac{\left(8.99 \times 10^9 \frac{\text{N} \cdot \text{m}^2}{\text{C}^2}\right)|3.00 \times 10^{-6} \text{ C}||2.00 \times 10^{-6} \text{ C}|}{(6.00 \times 10^{-2} \text{ m})^2}$$

$$+ \frac{\left(8.99 \times 10^9 \frac{\text{N} \cdot \text{m}^2}{\text{C}^2}\right)|6.00 \times 10^{-6} \text{ C}||2.00 \times 10^{-6} \text{ C}|(6.00 \times 10^{-2} \text{ m})}{[(7.00 \times 10^{-2} \text{ m})^2 + (6.00 \times 10^{-2} \text{ m})^2]^{3/2}}$$

$$= 23.24 \text{ N} = 23.2 \text{ N (rounded to 3 significant figures)}$$

Magnitude of net force on charge C:

$$F_{\text{net on C}} = \sqrt{(F_{\text{net on C},x})^2 + (F_{\text{net on C},y})^2}$$

$$F_{\text{net on C}} = \sqrt{(-9.636 \text{ N})^2 + (23.24 \text{ N})^2} = \boxed{25.2 \text{ N}}$$

Direction of net force on charge C:

$$\theta_C = \tan^{-1}\left(\frac{23.24 \text{ N}}{-9.636 \text{ N}}\right) = -67.5°$$

Since the force vector points in the second quadrant, the direction of the net force on charge C is $180° - 67.5° = \boxed{112.5°}$.

REFLECT

Be sure to consider the direction that each charge should move based on the sign of the charge. Since all of the charges are positive, the force between each charge is repulsive. However, this does not mean that each charge will move in the same direction.

16.63

SET UP

A uniformly charged plastic rod ($L = 0.100$ m) is sealed inside of a plastic bag. The net electric flux through the bag is $\Phi = 7.50 \times 10^5 \frac{\text{N} \cdot \text{m}^2}{\text{C}}$. We can use Gauss' law, $\Phi = \frac{q_{encl}}{\epsilon_0}$, to calculate the linear charge density of the rod.

SOLVE

$$\Phi = \frac{q_{encl}}{\epsilon_0} = \frac{\lambda L}{\epsilon_0}$$

$$\lambda = \frac{\Phi \epsilon_0}{L} = \frac{\left(7.50 \times 10^5 \frac{\text{N} \cdot \text{m}^2}{\text{C}}\right)\left(8.85 \times 10^{-12} \frac{\text{C}^2}{\text{N} \cdot \text{m}^2}\right)}{0.100 \text{ m}} = \boxed{6.64 \times 10^{-5} \frac{\text{C}}{\text{m}}}$$

REFLECT

The net electric flux through the bag is positive, which means the rod is positively charged. The exact shape of the bag is irrelevant because we are given the flux.

16.65

SET UP

A sphere carries a uniform surface charge density (charge per unit area) σ. We can use Gauss' law to find an expression for the electric field just outside the surface of the sphere. At this location, the Gaussian sphere will have approximately the same surface area as the charged sphere, which we'll call A. Assuming σ is positive, the electric field will point radially outward from the sphere.

SOLVE

$$AE_r = \frac{q_{encl}}{\epsilon_0}$$

$$AE_r = \frac{(\sigma A)}{\epsilon_0}$$

$$\boxed{E_r = \frac{\sigma}{\epsilon_0}, \text{ pointing radially outward}}$$

10 Chapter 16 Electrostatics I: Electric Charge, Forces, and Fields

REFLECT
The electric field is constant just outside the sphere.

Get Help: P'Cast 16.18 – Gauss' Law and Surface Charge

16.73

SET UP
A red blood cell carries an excess charge of $Q = -2.5 \times 10^{-12}$ C distributed uniformly over its surface. We will model the cells as spheres of diameter $D = 7.5 \times 10^{-6}$ m and mass $m_{cell} = 9.0 \times 10^{-14}$ kg. The number of excess electrons on the red blood cell is equal to Q divided by the charge on one electron, -1.602×10^{-19} C. In order to determine whether these extra electrons appreciably affect the mass of the cell, we can calculate the ratio of the mass of the excess electrons to the mass of the sphere; the mass of an electron is $m_{electron} = 9.11 \times 10^{-31}$ kg. Finally, the surface charge density of the red blood cell is equal to the total charge divided by the surface area of the cell.

SOLVE

Part a)

$$-2.5 \times 10^{-12} \text{ C} \times \frac{1 \text{ electron}}{-1.602 \times 10^{-19} \text{ C}} = \boxed{1.6 \times 10^7 \text{ electrons}}$$

Part b)

$$\frac{m_{electrons}}{m_{cell}} = \frac{(1.6 \times 10^7)(9.11 \times 10^{-31} \text{ kg})}{9.0 \times 10^{-14} \text{ kg}} = \boxed{1.6 \times 10^{-10}}$$

The extra mass is $\boxed{\text{not significant}}$.

Part c)

$$\sigma = \frac{Q}{A} = \frac{Q}{4\pi R^2} = \frac{Q}{4\pi \left(\frac{D}{2}\right)^2} = \frac{Q}{\pi D^2} = \frac{-2.5 \times 10^{-12} \text{ C}}{\pi (7.5 \times 10^{-6} \text{ m})^2}$$

$$= \boxed{-1.4 \times 10^{-2} \frac{\text{C}}{\text{m}^2}} \times \frac{1 \text{ electron}}{-1.602 \times 10^{-19} \text{ C}} = \boxed{8.8 \times 10^{16} \frac{\text{electrons}}{\text{m}^2}}$$

REFLECT
There would need to be about 10^{17} excess electrons, or an excess charge of around 10^{-2} C, for the mass of the excess electrons to be approximately equal to the mass of the cell.

Get Help: P'Cast 16.18 – Gauss' Law and Surface Charge

16.77

SET UP
The elephant nose fish can detect changes in an electric field as small as 3.0×10^{-6} N/C. We can set this equal to the expression for the electric field due to a point charge, $E = \frac{k|Q|}{r^2}$, in order to calculate the minimum charge required to create a field of this strength at a distance

of $r = 0.75$ m. This charge divided by the magnitude of the charge on one electron (1.602×10^{-19} C) is equal to the number of electrons required to achieve this charge.

SOLVE

Part a)

$$E = \frac{k|Q|}{r^2}$$

$$|Q| = \frac{Er^2}{k} = \frac{\left(3.0 \times 10^{-6} \frac{\text{N}}{\text{C}}\right)(0.75 \text{ m})^2}{\left(8.99 \times 10^9 \frac{\text{N} \cdot \text{m}^2}{\text{C}^2}\right)} = \boxed{1.9 \times 10^{-16} \text{ C}}$$

Part b)

$$1.9 \times 10^{-16} \text{ C} \times \frac{1 \text{ electron}}{1.602 \times 10^{-19} \text{ C}} = \boxed{1.2 \times 10^3 \text{ electrons}}$$

REFLECT

The fish can only detect *changes* in the local electric field of 3.0×10^{-6} N/C, so the additional field can either add to or subtract from the unperturbed field.

Get Help: Picture It – Electric Field
Interactive Example – Particle Beam
P'Cast 16.2 – Calculating Electric Force
P'Cast 16.6 – Where Is the Electric Field Zero?

16.79

SET UP

Six positive charges ($q_1 = 1 \times 10^{-3}$ C, $q_2 = 2 \times 10^{-3}$ C, $q_3 = 3 \times 10^{-3}$ C, $q_4 = 4 \times 10^{-3}$ C, $q_5 = 5 \times 10^{-3}$ C, $q_6 = 6 \times 10^{-3}$ C) are arranged in a regular hexagon with 5.00-cm-long sides (see figure). The electric field at the center of the hexagon is equal to the vector sum of the electric fields due to each point charge. The magnitude of the electric field due to a point charge Q at a distance r away is $E = \frac{k|Q|}{r^2}$. Since all of the charges are positive, the electric field will point away from each charge toward the center of the hexagon. Note that the y component of the electric field due to charges q_3 and q_6 will be zero.

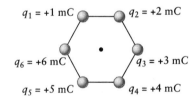

Figure 16-3 Problem 79

SOLVE

Magnitudes of the field due to each point charge at the center:

$$E_1 = \frac{k|q|_1}{r_1^2} = \frac{\left(8.99 \times 10^9 \frac{\text{N} \cdot \text{m}^2}{\text{C}^2}\right)(1 \times 10^{-3} \text{ C})}{(5.00 \times 10^{-2} \text{ m})^2} = 3.596 \times 10^9 \frac{\text{N}}{\text{C}}$$

$$E_2 = \frac{k|q_2|}{r_2^2} = \frac{k(2|q_1|)}{r_1^2} = 2E_1$$

12   Chapter 16 Electrostatics I: Electric Charge, Forces, and Fields

$$E_3 = \frac{k|q_3|}{r_3^2} = \frac{k(3|q_1|)}{r_1^2} = 3E_1$$

$$E_4 = \frac{k|q_4|}{r_4^2} = \frac{k(4|q_1|)}{r_1^2} = 4E_1$$

$$E_5 = \frac{k|q_5|}{r_5^2} = \frac{k(5|q_1|)}{r_1^2} = 5E_1$$

$$E_6 = \frac{k|q_6|}{r_6^2} = \frac{k(6|q_1|)}{r_1^2} = 6E_1$$

x component of the field at the center:

$$E_x = E_1\cos(60°) - E_2\cos(60°) - E_3 - E_4\cos(60°) + E_5\cos(60°) + E_6$$

$$= [E_1 - E_2 - E_4 + E_5]\cos(60°) + E_6 - E_3 = [E_1 - 2E_1 - 4E_1 + 5E_1]\cos(60°) + 6E_1 - 3E_1$$

$$= [1 - 2 - 4 + 5]E_1\cos(60°) + 6E_1 - 3E_1 = 3E_1 = 3\left(3.596 \times 10^9 \frac{\text{N}}{\text{C}}\right) = \boxed{1.08 \times 10^{10}\frac{\text{N}}{\text{C}}}$$

y component of the field at the center:

$$E_y = -E_1\sin(60°) - E_2\sin(60°) + E_4\sin(60°) + E_5\sin(60°)$$

$$= [-E_1 - E_2 + E_4 + E_5]\sin(60°) = [-E_1 - 2E_1 + 4E_1 + 5E_1]\sin(60°)$$

$$= [-1 - 2 + 4 + 5]E_1\sin(60°) = \frac{6\sqrt{3}}{2}E_1 = 3\sqrt{3}\left(3.596 \times 10^9 \frac{\text{N}}{\text{C}}\right) = \boxed{1.87 \times 10^{10}\frac{\text{N}}{\text{C}}}$$

REFLECT
There is more positive charge to the left of the center than the right and below than above, so the electric field at the center should point up and to the right. The field must point at an angle of 60 degrees above the horizontal due to symmetry.

Get Help: Picture It – Electric Field
Interactive Example – Particle Beam
P'Cast 16.2 – Calculating Electric Force
P'Cast 16.3 – Three Charges in a Line
P'Cast 16.6 – Where Is the Electric Field Zero?

16.83

SET UP
The nucleus of an iron atom contains 26 protons and has a radius of $r_{\text{nucleus}} = 4.6 \times 10^{-15}$ m. The radius of the atom is 0.50×10^{-10} m. We can treat the nucleus as a positive point charge of magnitude $26e$ for all points outside of the nucleus. The magnitude of the electric field due to a point charge is $E = \frac{k|Q|}{r^2}$; the field will point radially outward because the nucleus

is positively charged. We'll assume that the net force acting on the outermost electron is equal to the force due to the electric field of the nucleus at that point. We can calculate that electron's acceleration using Newton's second law.

SOLVE
Part a)

$$E = \frac{k|Q|}{r^2} = \frac{k(26|e|)}{r_{nucleus}^2} = \frac{\left(8.99 \times 10^9 \frac{N \cdot m^2}{C^2}\right)(26)|1.602 \times 10^{-19} \, C|}{(4.6 \times 10^{-15} \, m)^2} = 1.8 \times 10^{21} \frac{N}{C}$$

The electric field has a magnitude of $\boxed{1.8 \times 10^{21} \frac{N}{C} \text{ and points radially outward}}$.

Part b)

$$E = \frac{k|Q|}{r^2} = \frac{k(26|e|)}{r_{atom}^2} = \frac{\left(8.99 \times 10^9 \frac{N \cdot m^2}{C^2}\right)(26)|1.602 \times 10^{-19} \, C|}{(0.5 \times 10^{-10} \, m)^2} = 1.5 \times 10^{13} \frac{N}{C}$$

The electric field has a magnitude of $\boxed{1.5 \times 10^{13} \frac{N}{C} \text{ and points radially outward}}$.

Part c)

$$\sum F_{ext} = F_{electric} = |q|E = m_{electron} a$$

$$a = \frac{|q|E}{m_{electron}} = \frac{|-1.602 \times 10^{-19} \, C|\left(1.5 \times 10^{13} \frac{N}{C}\right)}{9.11 \times 10^{-31} \, kg} = 2.6 \times 10^{24} \frac{m}{s^2}$$

The acceleration of the outermost electron has a magnitude of $\boxed{2.6 \times 10^{24} \frac{m}{s^2} \text{ and}}$ $\boxed{\text{points radially inward}}$.

REFLECT
The force acting on the outermost electron will be much less than this due to the effects of the other 25 electrons. These electrons are closer to the nucleus and "screen" the charge of the nucleus so the effective field at the position of the outermost electron is smaller.

Get Help: Picture It – Electric Field
Interactive Example – Particle Beam
P'Cast 16.2 – Calculating Electric Force
P'Cast 16.3 – Three Charges in a Line
P'Cast 16.6 – Where Is the Electric Field Zero?

16.87

SET UP

Two hollow, concentric, spherical shells are covered with charge. The inner sphere has a radius R_i and a surface charge density of $+\sigma_i$; the outer sphere has a radius R_o and a surface charge density of $-\sigma_o$. We can use Gauss' law to derive an expression for the electric field everywhere in space. We will use a spherical Gaussian surface of radius r because of the symmetry of the charge distribution. Since the spheres are hollow, the charge only exists on the surface of each sphere.

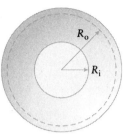

Figure 16-4 Problem 87

SOLVE

Part a)

$r < R_i$

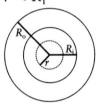

$$\Phi = \frac{q_{encl}}{\varepsilon_0} = \frac{0}{\varepsilon_0}$$

$$\boxed{\vec{E} = 0}$$

Part b)

$R_i < r < R_o$

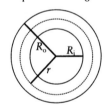

$$\Phi = \frac{q_{encl}}{\varepsilon_0}$$

$$E_r A = \frac{q_{encl}}{\varepsilon_0}$$

$$E_r = \frac{q_{encl}}{A\varepsilon_0} = \frac{4\pi R_i^2 \sigma_i}{(4\pi r^2)\varepsilon_0} = \frac{R_i^2 \sigma_i}{r^2 \varepsilon_0}$$

The electric field has a magnitude of $\boxed{\dfrac{R_i^2 \sigma_i}{r^2 \varepsilon_0} \text{ and points radially outward}}$.

Part c)

$r > R_o$

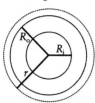

$$\Phi = \frac{q_{encl}}{\varepsilon_0}$$

$$E_r A = \frac{q_{encl}}{\varepsilon_0}$$

$$E_r = \frac{q_{encl}}{A\varepsilon_0} = \frac{4\pi R_i^2 \sigma_i - 4\pi R_o^2 \sigma_o}{(4\pi r^2)\varepsilon_0} = \frac{R_i^2 \sigma_i - R_o^2 \sigma_o}{r^2 \varepsilon_0}$$

The magnitude of the electric field is $\boxed{\dfrac{R_i^2 \sigma_i - R_o^2 \sigma_o}{r^2 \varepsilon_0}}$ and points radially outward if $R_i^2 \sigma_i > R_o^2 \sigma_o$ $\boxed{\text{or inward if } R_i^2 \sigma_i < R_o^2 \sigma_o}$.

REFLECT

The electric field in the region between the shells can either point radially inward or radially outward depending upon the relative sizes and the relative surface charge densities of the shells.

Get Help: P'Cast 16.18 – Gauss' Law and Surface Charge

Chapter 17
Electrostatics II: Electric Potential Energy and Electric Potential

Conceptual Questions

17.5 This statement makes sense only if the zero point of the electric potential has been previously defined.

>   **Get Help:** Picture It – Electric Potential
>   P'Cast 17.4 – Electric Potential Difference in a Uniform Field I
>   P'Cast 17.5 – Electric Potential Difference in a Uniform Field II
>   P'Cast 17.6 – Transmission Electron Microscope

17.9 Part a) Yes, a region of constant potential must have zero electric field.

Part b) No, if the electric field is zero, the potential need only be constant.

17.15 The charge on the capacitor remains constant, while the capacitance decreases. Therefore, the energy stored in the capacitor increases.

>   **Get Help:** P'Cast 17.9 – A Defibrillator Capacitor

Multiple-Choice Questions

17.21 D (its potential energy decreases and its electric potential decreases). The electric field points in the direction of the force a positive charge would experience. A positive charge moving in this direction would lower its potential energy. The electric field also points toward regions of lower potential.

>   **Get Help:** Picture It – Electric Potential
>   P'Cast 17.4 – Electric Potential Difference in a Uniform Field I
>   P'Cast 17.5 – Electric Potential Difference in a Uniform Field II
>   P'Cast 17.6 – Transmission Electron Microscope

17.27 B (doubled).

$$\frac{U_{electric,2}}{U_{electric,1}} = \frac{\left(\frac{1}{2}\frac{q_2^2}{C_2}\right)}{\left(\frac{1}{2}\frac{q_1^2}{C_1}\right)} = \frac{\left(\frac{q^2}{C_2}\right)}{\left(\frac{q^2}{C_1}\right)} = \frac{C_1}{C_2} = \frac{\left(\frac{\varepsilon_0 A}{d_1}\right)}{\left(\frac{\varepsilon_0 A}{d_2}\right)} = \frac{d_2}{d_1} = \frac{2d_1}{d_1} = \boxed{2}$$

>   **Get Help:** Picture It – Capacitance
>   P'Cast 17.7 – A Parallel-Plate Capacitor
>   P'Cast 17.8 – Insulin Release

Estimation/Numerical Analysis

17.33 The electric potential of an electron located a distance of approximately 5×10^{-11} m from the nucleus of a hydrogen atom is around 30 V.

> **Get Help:** Picture It – Electric Potential
> P'Cast 17.4 – Electric Potential Difference in a Uniform Field I
> P'Cast 17.5 – Electric Potential Difference in a Uniform Field II
> P'Cast 17.6 – Transmission Electron Microscope

Problems

17.35

SET UP

A uniform electric field of $E = 2.00 \times 10^3$ N/C points toward $+x$. A positive test charge $q_0 = +2.00 \times 10^{-9}$ C moves from $x_a = -0.300$ m to $x_b = 0.500$ m. The change in electric potential energy of a charge in an electric field is given by the equation $\Delta U_{electric} = -qEd\cos\theta$. Because the charge moves in the same direction as the electric field, the angle θ is 0. Knowing this test charge is released from rest at point a, we can use conservation of energy to calculate the kinetic energy of the charge when it passes through point b. If the charge were negative rather than positive, the negative charge would be accelerated toward $-x$ by the electric field and would not pass through point b.

SOLVE

Part a)

$$\Delta U_{electric} = -qEd\cos\theta = U_{electric,b} - U_{electric,a}$$

$$= -q_0 E(x_b - x_a)(\cos 0) = -q_0 E(x_b - x_a)$$

$$= -(2.00 \times 10^{-9}\,\text{C})\left(2.00 \times 10^3 \frac{\text{N}}{\text{C}}\right)[0.500\,\text{m} - (-0.300\,\text{m})]$$

$$= \boxed{-3.20 \times 10^{-6}\,\text{J}}$$

Part b)

$$U_{electric,i} + K_i = U_{electric,f} + K_f$$

$$\Delta U_{electric} = K_a - K_b = 0 - K_b$$

$$K_b = -\Delta U_{electric} = \boxed{3.2 \times 10^{-6}\,\text{J}}$$

Part c) If a negative charge were placed at rest at point a, then it would never reach point b unless an external force acted upon it. The charge would accelerate in the $-x$ direction away from point b.

18 Chapter 17 Electrostatics II: Electric Potential Energy and Electric Potential

REFLECT

Use your intuition to help keep your signs straight. Electrons naturally accelerate in the direction opposite to the electric field. Since they will accelerate from point a toward point b, point b must be at a lower potential energy.

> **Get Help:** Picture It – Electrostatic Energy
> Interactive Example – Three Charges
> P'Cast 17.1 – Electric Potential Energy Difference in a Uniform Field
> P'Cast 17.2 – Electric Potential Energy and Nuclear Fission
> P'Cast 17.3 – Electric Potential Energy of Three Charges

17.39

SET UP

A charge ($q_0 = +2.0$ C) is moved through a potential difference of $\Delta V = +9.0$ V. Assuming the charge starts and ends at rest, the work required to move the charge is: $W_{required} = -W_{electric} = \Delta U_{electric}$. Also, recall that the change in potential is defined as $\Delta V = \dfrac{\Delta U_{electric}}{q_0}$.

SOLVE

$$\Delta V = \frac{\Delta U_{electric}}{q_0} = \frac{W_{required}}{q_0}$$

$$W_{required} = q_0 \Delta V = (2.0 \text{ C})(9.0 \text{ V}) = \boxed{18 \text{ J}}$$

REFLECT

It should require energy to move a positive charge to an area of higher potential.

> **Get Help:** Picture It – Electric Potential
> P'Cast 17.4 – Electric Potential Difference in a Uniform Field I
> P'Cast 17.5 – Electric Potential Difference in a Uniform Field II
> P'Cast 17.6 – Transmission Electron Microscope

17.45

SET UP

Two point charges are placed on the x-axis: $q_1 = 0.500$ μC is located at $x = 0$ and $q_2 = -0.200$ μC is located at $x = 10.0$ cm. The electric potential due to a point charge Q as a function of distance is $V(r) = \dfrac{kQ}{r}$. We can calculate the position where the electric potential is equal to zero by adding the expressions for the potential due to each charge, setting the sum to zero, and solving for x.

SOLVE

$$V(r) = \frac{kQ}{r}$$

$$V(x) = 0 = \frac{kq_1}{x} + \frac{kq_2}{(x - 10.0 \text{ cm})} \quad (x \text{ in cm})$$

Chapter 17 Electrostatics II: Electric Potential Energy and Electric Potential

$$-\frac{q_1}{x} = \frac{q_2}{(x - 10.0 \text{ cm})}$$

$$x = \frac{(10.0 \text{ cm})q_1}{q_1 + q_2} = \frac{(10.0 \text{ cm})(0.500 \times 10^{-6} \text{ C})}{(0.500 \times 10^{-6} \text{ C}) + (-0.200 \times 10^{-6} \text{ C})} = \boxed{16.7 \text{ cm}}$$

REFLECT

It makes sense that the position where $V = 0$ is to the right of both charges since the magnitude of q_1 is larger than the magnitude of q_2, q_1 is positive, and q_2 is negative.

Get Help: Picture It – Electric Potential
P'Cast 17.4 – Electric Potential Difference in a Uniform Field I
P'Cast 17.5 – Electric Potential Difference in a Uniform Field II
P'Cast 17.6 – Transmission Electron Microscope

17.51

SET UP

We are asked to draw the equipotential lines and electric field lines for two pairs of charges—two $+q$ and two $-q$. The equipotential lines will be the closest together near the charges and spread out as we move away. Electric field lines start on positive charges, end on negative charges, and cannot cross one another.

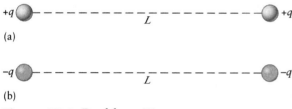

Figure 17-1 Problem 51

SOLVE

Parts a) (Equipotential lines) and b) (Electric fields)

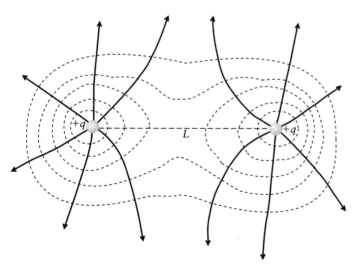

Figure 17-2 Problem 51

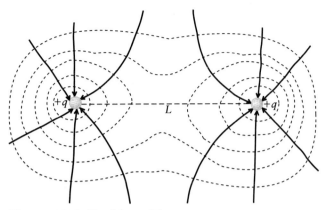

Figure 17-3 Problem 51

REFLECT

Although the shapes of the equipotential curves are the same in both setups, the potentials those lines correspond to would not be the same.

17.59

SET UP

A parallel plate capacitor has square plates ($L = 1.00$ m). We can use the expression for the capacitance of a parallel plate capacitor, $C = \dfrac{\varepsilon_0 A}{d}$, to calculate the separation distance d required to give $C = 8850 \times 10^{-12}$ F.

SOLVE

$$C = \frac{\varepsilon_0 A}{d} = \frac{\varepsilon_0 L^2}{d}$$

$$d = \frac{\varepsilon_0 L^2}{C} = \frac{\left(8.85 \times 10^{-12} \dfrac{\text{F}}{\text{m}}\right)(1.00 \text{ m})^2}{8850 \times 10^{-12} \text{ F}} = 0.00100 \text{ m} = \boxed{1.00 \text{ mm}}$$

REFLECT

The dielectric constant of air is $\kappa = 1.00058$, which is approximately 1.

Get Help: Picture It – Capacitance
P'Cast 17.7 – A Parallel-Plate Capacitor
P'Cast 17.8 – Insulin Release

17.65

SET UP

The capacitor ($C = 20.0 \times 10^{-6}$ F) in a defibrillator has a voltage of 10.0×10^3 V. The energy released into the patient is equal to the potential energy initially stored by the capacitor, $U_{\text{electric}} = \dfrac{1}{2}CV^2$.

SOLVE

$$U_{\text{electric}} = \frac{1}{2}CV^2 = \frac{1}{2}(20.0 \times 10^{-6} \text{ F})(10.0 \times 10^3 \text{ V})^2 = \boxed{1.00 \times 10^3 \text{ J}}$$

REFLECT
We assumed that all of the energy stored in the capacitor is released into the patient.

Get Help: P'Cast 17.9 – A Defibrillator Capacitor

17.69

SET UP

Two capacitors, C_1 and C_2, are arranged in series and in parallel. The equivalent series capacitance is $C_{series} = 2.00 \ \mu\text{F}$, and the equivalent parallel capacitance is $C_{parallel} = 8.00 \ \mu\text{F}$. We can solve for the values of C_1 and C_2 using the expressions for the equivalent series $\left(\dfrac{1}{C_{equiv,series}} = \dfrac{1}{C_1} + \dfrac{1}{C_2}\right)$ and parallel ($C_{equiv,parallel} = C_1 + C_2$) capacitances.

SOLVE

Equivalent capacitances:

$$\frac{1}{C_{series}} = \frac{1}{C_1} + \frac{1}{C_2}$$

$$C_{series} = \frac{C_1 C_2}{C_1 + C_2}$$

$$C_{parallel} = C_1 + C_2$$

Solving for C_1 and C_2:

$$C_{series}(C_1 + C_2) = C_1 C_2$$

$$C_{series} C_{parallel} = (C_{parallel} - C_2) C_2$$

$$C_2^2 - C_{parallel} C_2 + C_{series} C_{parallel} = 0$$

$$C_2 = \frac{-(-C_{parallel}) \pm \sqrt{(-C_{parallel})^2 - 4(1)(C_{series} C_{parallel})}}{2(1)} = \frac{C_{parallel} \pm \sqrt{C_{parallel}^2 - 4 C_{series} C_{parallel}}}{2}$$

$$= \frac{(8.00 \ \mu\text{F}) \pm \sqrt{(8.00 \ \mu\text{F})^2 - 4(2.00 \ \mu\text{F})(8.00 \ \mu\text{F})}}{2} = \frac{(8.00 \ \mu\text{F}) \pm 0}{2} = \boxed{4.00 \ \mu\text{F}}$$

$$C_1 = C_{parallel} - C_2 = (8.00 \ \mu\text{F}) - (4.00 \ \mu\text{F}) = \boxed{4.00 \ \mu\text{F}}$$

REFLECT

Quickly double-checking our answer, we find that $4 + 4 = 8$, and $(1/4) + (1/4) = (1/2)$, which matches the values listed in the problem statement.

Get Help: P'Cast 17.10 – Two Capacitors in Series or in Parallel
P'Cast 17.11 – Multiple Capacitors

22 Chapter 17 Electrostatics II: Electric Potential Energy and Electric Potential

17.75

SET UP

Three capacitors—$C_1 = 10.0\ \mu F$, $C_2 = 40.0\ \mu F$, and $C_3 = 100.0\ \mu F$—are connected in series across a 12.0-V battery. The equivalent capacitance of three capacitors in series is $\dfrac{1}{C_{equiv}} = \dfrac{1}{C_1} + \dfrac{1}{C_2} + \dfrac{1}{C_3}$. When they are wired up to the battery, the capacitors will store charge. The capacitors are wired in series, which means the charge on each capacitor will be the same. Because of this, we can use the equivalent capacitance of the capacitor network and the potential difference across the battery to calculate the charge on each capacitor. The potential difference $V = \dfrac{q}{C}$ across each capacitor will not be equal, though, since the capacitances are different.

SOLVE

Part a)

$$\frac{1}{C_{equiv}} = \frac{1}{C_1} + \frac{1}{C_2} + \frac{1}{C_3} = \frac{1}{10.0\ \mu F} + \frac{1}{40.0\ \mu F} + \frac{1}{100.0\ \mu F} = \frac{27}{200.0\ \mu F}$$

$$C_{equiv} = \frac{200.0\ \mu F}{27} = \boxed{7.41\ \mu F}$$

Part b)

$$q = C_{equiv} V = (7.41 \times 10^{-6}\ F)(12.0\ V) = 8.89 \times 10^{-5}\ C = \boxed{88.9\ \mu C}$$

Part c)

$$V_1 = \frac{q}{C_1} = \frac{88.9 \times 10^{-6}\ C}{10.0 \times 10^{-6}\ F} = \boxed{8.89\ V}$$

$$V_2 = \frac{q}{C_2} = \frac{88.9 \times 10^{-6}\ C}{40.0 \times 10^{-6}\ F} = \boxed{2.22\ V}$$

$$V_3 = \frac{q}{C_3} = \frac{88.9 \times 10^{-6}\ C}{100.0 \times 10^{-6}\ F} = \boxed{0.889\ V}$$

REFLECT

The sum of the potential differences across each capacitor should be 12.0 V: (8.89 V) + (2.22 V) + (0.889 V) = 12.0 V. The potential energy stored in the three capacitors in series is equal to the potential energy stored by the equivalent capacitor.

Get Help: P'Cast 17.10 – Two Capacitors in Series or in Parallel
P'Cast 17.11 – Multiple Capacitors

17.79

SET UP

An air-gap capacitor ($C_0 = 2800 \times 10^{-12}\ F$) accumulates a charge q_0 when it is connected to a 16-V battery. While the battery is still connected, a dielectric material ($\kappa = 5.8$) is placed between the plates of the capacitor, which increases the capacitance to $C = \kappa C_0$. The final charge on the capacitor is q. We can use the definition of capacitance to calculate the amount of charge that flowed from the battery to the capacitor upon adding the dielectric.

SOLVE

$$q = CV$$

$$\Delta q = q - q_0 = CV - C_0V = (\kappa C_0)V - C_0V = C_0V(\kappa - 1)$$

$$= (2800 \times 10^{-12} \text{ F})(16 \text{ V})(5.8 - 1) = \boxed{2.2 \times 10^{-7} \text{ C}}$$

REFLECT

Inserting a dielectric into a capacitor increases its capacitance, which means it can store more charge for a given potential difference. Therefore, charge will flow from the battery onto the plates of the capacitor. If the battery were not connected when the dielectric was inserted, the charge on the plates would remain constant and the potential difference across the capacitor would decrease.

Get Help: P'Cast 17.12 – Cell Membrane Capacitance

17.85

SET UP

Three particles, each with charge q, are located at different corners of a rhombus with sides of length a, a short diagonal of length a, and a long diagonal of length b. The relationship between the sides of the rhombus and its diagonals is: $4a^2 = a^2 + b^2$. The total electric potential energy of the charge distribution is equal to the sum of the pairwise potential energies of the charges, $U_{\text{electric}} = \dfrac{kq_A q_B}{r}$. The required external work to bring a fourth charge q from infinity to the fourth corner of the rhombus is equal to the change in the particle's mechanical energy. Since the particle starts and ends at rest, the change in the kinetic energy is equal to zero. The potential energy of the particle at infinity is defined to be zero. Finally, we can use our answers to parts (a) and (b) to calculate the total electric potential energy of the charge distribution consisting of the four charges.

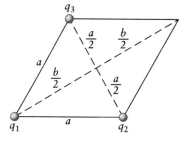

Figure 17-4 Problem 85

SOLVE

Part a)

$$U_{\text{electric}} = U_{q_1 q_2} + U_{q_2 q_3} + U_{q_1 q_3} = \frac{kq_1 q_2}{a} + \frac{kq_2 q_3}{a} + \frac{kq_1 q_3}{a} = \frac{k}{a}[q^2 + q^2 + q^2] = \boxed{\frac{3kq^2}{a}}$$

Part b)

$$-W_{\text{electric}} = W_{\text{required}} = \Delta U_{\text{electric}} = U_{\text{electric},f} - U_{\text{electric},i} = U_{\text{electric},f} - 0 = U_{q_1 q_4} + U_{q_2 q_4} + U_{q_3 q_4}$$

$$= \frac{kq_1 q_4}{b} + \frac{kq_2 q_4}{a} + \frac{kq_3 q_4}{a}$$

$$= \frac{kq^2}{b} + \frac{kq^2}{a} + \frac{kq^2}{a} = \frac{kq^2}{\sqrt{4a^2 - a^2}} + \frac{2kq^2}{a} = \boxed{\left(2 + \frac{1}{\sqrt{3}}\right)\frac{kq^2}{a}}$$

Part c)

$$U_{\text{electric}} = (U_{q_1q_2} + U_{q_2q_3} + U_{q_1q_3}) + (U_{q_1q_4} + U_{q_2q_4} + U_{q_3q_4}) = \left(\frac{3kq^2}{a}\right) + \left(\left(2 + \frac{1}{\sqrt{3}}\right)\frac{kq^2}{a}\right)$$

$$= \boxed{\left(5 + \frac{1}{\sqrt{3}}\right)\frac{kq^2}{a}}$$

REFLECT

We can talk about the electric potential due to one charge, but the electric potential energy due to the interaction between two charges.

> **Get Help:** Picture It – Electrostatic Energy
> Interactive Example – Three Charges
> P'Cast 17.1 – Electric Potential Energy Difference in a Uniform Field
> P'Cast 17.2 – Electric Potential Energy and Nuclear Fission
> P'Cast 17.3 – Electric Potential Energy of Three Charges

17.89

SET UP

We are asked to find the equivalent capacitance of four capacitors—$C_1 = 3$ pF, $C_2 = 6$ pF, $C_3 = 4$ pF, $C_4 = 6$ pF— wired up as shown (see figure). The potential difference over the entire top branch (made up of C_3 and C_4) is equal to the potential difference over the entire bottom branch (made up of C_1 and C_2), which means these branches are in parallel with one another. The capacitors in each branch (C_1 and C_2, C_3 and C_4) are wired in series since the current through each branch is the same throughout the branch. We can find the overall equivalent capacitance by first finding the equivalent capacitance of each branch and then the equivalent capacitance of the two branches together.

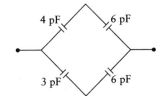

Figure 17-5 Problem 89

The equivalent capacitance of two capacitors in series is $\dfrac{1}{C_{\text{equiv}}} = \dfrac{1}{C_1} + \dfrac{1}{C_2}$; the equivalent capacitance of two capacitors in parallel is $C_{\text{equiv}} = C_1 + C_2$.

SOLVE

Redrawn diagram:

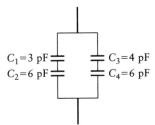

Figure 17-6 Problem 89

Equivalent capacitance:

$$\frac{1}{C_{\text{equiv}}} = \frac{1}{C_{\text{series}}} = \frac{1}{C_{12}} = \frac{1}{C_1} + \frac{1}{C_2} = \frac{1}{3 \text{ pF}} + \frac{1}{6 \text{ pF}} = \frac{3}{6 \text{ pF}}$$

$$C_{12} = 2 \text{ pF}$$

$$\frac{1}{C_{equiv}} = \frac{1}{C_{series}} = \frac{1}{C_{34}} = \frac{1}{C_3} + \frac{1}{C_4} = \frac{1}{4 \text{ pF}} + \frac{1}{6 \text{ pF}} = \frac{5}{12 \text{ pF}}$$

$$C_{34} = \frac{12}{5} \text{pF}$$

$$C_{equiv} = C_{parallel} = C_{1234} = C_{12} + C_{34} = (2 \text{ pF}) + \left(\frac{12}{5}\text{pF}\right)$$

$$= \boxed{\frac{22}{5} \text{ pF} = 4.4 \text{ pF} \approx 4 \text{ pF} \text{ (rounded to one significant figure)}}$$

REFLECT

Redrawing the circuit diagram using right angles helps when determining which elements are in series and which are in parallel.

Get Help: P'Cast 17.10 – Two Capacitors in Series or in Parallel
P'Cast 17.11 – Multiple Capacitors

17.99

SET UP

A parallel plate capacitor has an area A and separation d. A conducting slab of thickness d' is inserted between, and parallel to, the plates. This set up is equivalent to two parallel plate capacitors in series. If we define the distances of the gaps between the plates as s_1 and s_2, the separation distances for each "sub-capacitor" are $(d - s_1 - d')$ and $(d - s_2 - d')$, where $d = s_1 + s_2 + d'$. We can use the general expression for the capacitance of a parallel plate capacitor, $C = \frac{\varepsilon_0 A}{d}$, and the equivalent capacitance of two capacitors in series, $\frac{1}{C_{equiv}} = \frac{1}{C_1} + \frac{1}{C_2}$, to calculate the capacitance of the object upon adding the conducting slab.

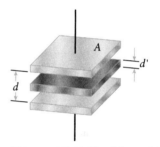

Figure 17-7 Problem 99

SOLVE

Part a)

Original capacitance:

$$C_i = \frac{\varepsilon_0 A}{d}$$

Adding the conducting slab:

$$\frac{1}{C_{equiv}} = \frac{1}{C_f} = \frac{1}{C_1} + \frac{1}{C_2} = \frac{1}{\left(\frac{\varepsilon_0 A}{d - s_1 - d'}\right)} + \frac{1}{\left(\frac{\varepsilon_0 A}{d - s_2 - d'}\right)} = \frac{d - s_1 - d' + d - s_2 - d'}{\varepsilon_0 A}$$

$$= \frac{2d - d' - (s_1 + s_2 + d')}{\varepsilon_0 A} = \frac{2d - d' - (d)}{\varepsilon_0 A} = \frac{d - d'}{\varepsilon_0 A}$$

$$\boxed{C_\text{f} = \frac{\varepsilon_0 A}{d - d'}}$$

The capacitance has increased upon adding the conducting slab.

Part b) Since our expression is independent of s_1 and s_2, the effect is independent of the location of the slab.

REFLECT

Even though we treated the "sub-capacitors" as being in series, the capacitance has *increased* upon adding the conducting slab. This may seem counterintuitive—adding identical capacitors in series lowers their equivalent capacitance—but, keep in mind, the "sub-capacitors" are *not identical to the original capacitor*! These "sub-capacitors" have the same cross-sectional area, but a *smaller* separation distance than the original capacitor.

Get Help: Picture It – Capacitance
P'Cast 17.7 – A Parallel-Plate Capacitor
P'Cast 17.8 – Insulin Release

Chapter 18
Electric Charges in Motion

Conceptual Questions

18.3 There is no contradiction. If a conductor is in electrostatic equilibrium, the electric field with it must be zero. However, any conductor that is carrying a current is definitely not in electrostatic equilibrium.

>   **Get Help:** P'Cast 18.1 – Charging a Sphere
>   P'Cast 18.2 – Electron Drift Speed in a Flashlight

18.7 A small resistance will dissipate more power and generate more heat.

>   **Get Help:** P'Cast 18.10 – Power in Series and Parallel Circuits

18.11 The voltage drop across each bulb in the old series string was about 1/50 of 110 V, or 2.2 V. The modern parallel connection puts the full 110 V across each bulb. Placing 110 V across one of the old bulbs, designed to operate at 2.2 V, would result in excessive current in the filament, which would burn out the bulb immediately, perhaps in a spectacular manner.

>   **Get Help:** Picture It – Kirchoff's Rules
>   P'Cast 18.7 – Giant Axons in Squid Revisited
>   P'Cast 18.9 – Resistors in Combination

Multiple-Choice Questions

18.15 **A** (increases along length of the wire). The drift speed is inversely proportional to the cross-sectional area of the wire.

>   **Get Help:** P'Cast 18.1 – Charging a Sphere
>   P'Cast 18.2 – Electron Drift Speed in a Flashlight

18.19 **D** (2 A). Since we are using the same wire, the resistance remains constant:

$$\frac{V_2}{V_1} = \frac{i_2 R}{i_1 R}$$

$$i_2 = \left(\frac{V_2}{V_1}\right) i_1 = \left(\frac{2\text{ V}}{1\text{ V}}\right)(1\text{ A}) = 2\text{ A}$$

>   **Get Help:** P'Cast 18.3 – Stretching a Wire
>   P'Cast 18.4 – Calculating Current

18.23 B (decreases). A light bulb acts as a resistor, and the equivalent resistance for two resistors in parallel is $\dfrac{1}{R_{equiv}} = \dfrac{1}{R_1} + \dfrac{1}{R_2}$.

Get Help: Picture It – Kirchoff's Rules
P'Cast 18.7 – Giant Axons in Squid Revisited
P'Cast 18.9 – Resistors in Combination

Estimation/Numerical Analysis

18.31 The resistor is about 1 kΩ and the capacitor is about 100 μF. This gives a time constant of $RC = (1\ \text{k}\Omega)(100\ \mu\text{F}) = 0.1$ s, which is on the smaller side.

Get Help: Interactive Example – RC I
Interactive Example – RC II
Interactive Example – RC III

Problems

18.35

SET UP

A synchrotron facility creates an electron beam with a current $i = 0.487$ A. We can calculate the number of electrons that pass a given point in an hour from the current and the magnitude of the charge on an electron, 1.60×10^{-19} C. Note that one ampere is equivalent to one coulomb/second, so the electric current can be expressed as 0.487 C/s.

SOLVE

$$1\ \text{h} \times \frac{3600\ \text{s}}{1\ \text{h}} \times \frac{0.487\ \text{C}}{1\ \text{s}} \times \frac{1\ \text{electron}}{1.60 \times 10^{-19}\ \text{C}} = \boxed{1.10 \times 10^{22}\ \text{electrons}}$$

REFLECT

We were given the current, so we didn't need to use the dimensions of the synchrotron or the speed of the electrons.

Get Help: P'Cast 18.1 – Charging a Sphere
P'Cast 18.2 – Electron Drift Speed in a Flashlight

18.41

SET UP

A length of wire has an initial length $L_1 = 8.00$ m and resistance $R_1 = 4.00\ \Omega$. The wire is then stretched uniformly to a final length $L_2 = 16.0$ m. Since the mass and density of the wire must be constant, the volume of the wire must also remain constant. Therefore, the cross-sectional area will decrease upon stretching. We can then calculate the new resistance of the wire R_2 using the definition of R in terms of the resistivity, $R = \dfrac{\rho L}{A}$.

SOLVE

New cross-sectional area:

$$A_1 L_1 = A_2 L_2$$

$$A_1(8.00 \text{ m}) = A_2(16.0 \text{ m})$$

$$A_2 = \frac{A_1}{2}$$

New resistance:

$$\frac{R_2}{R_1} = \frac{\left(\frac{\rho L_2}{A_2}\right)}{\left(\frac{\rho L_1}{A_1}\right)} = \frac{L_2 A_1}{A_2 L_1} = \frac{L_2 A_1}{\left(\frac{A_1}{2}\right) L_1} = \frac{2 L_2}{L_1} = \frac{2(16.0 \text{ m})}{8.00 \text{ m}} = 4.00$$

$$R_2 = 4 R_1 = (4.00)(4.00 \text{ }\Omega) = \boxed{16.0 \text{ }\Omega}$$

REFLECT

Looking at the expression for the resistance in terms of the resistivity, $R = \frac{\rho L}{A}$, increasing the length of the wire and decreasing the cross-sectional area both cause the resistance to increase.

Get Help: P'Cast 18.3 – Stretching a Wire
P'Cast 18.4 – Calculating Current

18.47

SET UP

Electrical cables are made up of a large number of strands of conducting wire. Each cable can have anywhere between 850 to 950 strands, each with a diameter of 0.72 ± 0.07 mm. The resistance of the cable is equal to the sum of the resistances of each component strand. We can determine an expression for this total resistance from combining the expression for the resistance in terms of the resistivity with Ohm's law. Plugging in the provided values will allow us to calculate the ratio of the lowest possible resistance to the average resistance and the ratio of the highest possible resistance to the average resistance of one of these cables.

SOLVE

$$i_{\text{total}} = n i_{\text{strand}} = n\left(\frac{V}{R_{\text{strand}}}\right) = \frac{nV}{\left(\frac{\rho L}{A}\right)} = \frac{nV(\pi r^2)}{\rho L} = \frac{n\pi V\left(\frac{d}{2}\right)^2}{\rho L} = \frac{n\pi V d^2}{4\rho L}$$

$$i_{\text{total}} = \frac{V}{R_{\text{total}}} = \frac{n\pi V d^2}{4\rho L}$$

$$R_{\text{total}} = \frac{4\rho L}{n\pi d^2}$$

Part a)

$$\frac{R_{low}}{R_{avg}} = \frac{\left(\dfrac{4\rho L}{n_{low}\pi d_{low}^2}\right)}{\left(\dfrac{4\rho L}{n_{avg}\pi d_{avg}^2}\right)} = \frac{n_{avg}d_{avg}^2}{n_{low}d_{low}^2} = \frac{(900)(0.72 \text{ mm})^2}{(950)(0.79 \text{ mm})^2} = \boxed{0.79}$$

Part b)

$$\frac{R_{high}}{R_{avg}} = \frac{\left(\dfrac{4\rho L}{n_{high}\pi d_{high}^2}\right)}{\left(\dfrac{4\rho L}{n_{avg}\pi d_{avg}^2}\right)} = \frac{n_{avg}d_{avg}^2}{n_{high}d_{high}^2} = \frac{(900)(0.72 \text{ mm})^2}{(850)(0.65 \text{ mm})^2} = \boxed{1.3}$$

REFLECT

The larger the diameter is for a wire, the smaller its resistance will be.

Get Help: P'Cast 18.5 – Resistance of a Potassium Channel

18.57

SET UP

Two resistors, $R_9 = 9.00 \text{ }\Omega$ and $R_3 = 3.00 \text{ }\Omega$, are wired in series across a battery ($\varepsilon = 9.00$ V). Since the resistors are connected together in series, the current through each of them is the same. We can calculate the current from the battery voltage and the equivalent resistance of the two resistors in series. The equivalent resistance for the resistors in series is $R_{equiv} = R_9 + R_3$. Even though the current is the same through both resistors, the voltage drop across each one will be different. We can calculate the voltage drops V_9 and V_3 from the current found in part (a) and from the individual resistances.

SOLVE

Part a)

Equivalent resistance:

$$R_{equiv} = R_9 + R_3 = (9.00 \text{ }\Omega) + (3.00 \text{ }\Omega) = 12.00 \text{ }\Omega$$

Current:

$$V = iR$$

$$\varepsilon = iR_{equiv}$$

$$i = \frac{\varepsilon}{R_{equiv}} = \frac{9.00 \text{ V}}{12.00 \text{ }\Omega} = \boxed{0.750 \text{ A}}$$

Part b)

$$V_9 = iR_9 = (0.750 \text{ A})(9.00 \text{ }\Omega) = \boxed{6.75 \text{ V}}$$

$$V_3 = iR_3 = (0.750 \text{ A})(3.00 \text{ }\Omega) = \boxed{2.25 \text{ V}}$$

REFLECT

The sum of the voltages across the two resistors must equal the voltage of the battery:

$$V_9 + V_3 \stackrel{?}{=} \varepsilon$$

$$(6.75 \text{ V}) + (2.25 \text{ V}) = 9.00 \text{ V}$$

Get Help: Picture It – Kirchoff's Rules
P'Cast 18.7 – Giant Axons in Squid Revisited
P'Cast 18.9 – Resistors in Combination

18.61

SET UP

Six resistors—$R_1 = 40\ \Omega$, $R_2 = 70\ \Omega$, $R_3 = 30\ \Omega$, $R_4 = 60\ \Omega$, $R_5 = 10\ \Omega$, and $R_6 = 20\ \Omega$—are wired together as shown in the figure. The equivalent resistance of this setup can be found from the expressions for the equivalent resistance for resistors in series and parallel, $\left(R_{equiv} = R_1 + R_2 + R_3 + \ldots + R_N \text{ and } \dfrac{1}{R_{equiv}} = \dfrac{1}{R_1} + \dfrac{1}{R_2} + \dfrac{1}{R_3} + \ldots + \dfrac{1}{R_N}, \text{ respectively} \right)$.

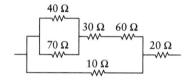

Figure 18-1 Problem 61

SOLVE

$$\frac{1}{R_{12}} = \frac{1}{R_1} + \frac{1}{R_2} = \frac{1}{40\ \Omega} + \frac{1}{70\ \Omega} = \frac{11}{280\ \Omega}$$

$$R_{12} = \frac{280}{11}\Omega$$

$$R_{1234} = R_{12} + R_3 + R_4 = \left(\frac{280}{11}\Omega\right) + (30\ \Omega) + (60\ \Omega) = \frac{1270}{11}\Omega$$

$$\frac{1}{R_{12345}} = \frac{1}{R_{1234}} + \frac{1}{R_5} = \frac{11}{1270\ \Omega} + \frac{1}{10\ \Omega} = \frac{138}{1270\ \Omega}$$

$$R_{12345} = \frac{1270}{138}\Omega$$

$$R_{equiv} = R_{12345} + R_6 = \left(\frac{1270}{138}\Omega\right) + (20\ \Omega) = \boxed{\frac{4030}{138}\Omega \approx 30\ \Omega}$$

REFLECT

When calculating the equivalent resistance of a resistor network, go step-by-step and start with the smallest unit.

Get Help: Picture It – Kirchoff's Rules
P'Cast 18.7 – Giant Axons in Squid Revisited
P'Cast 18.9 – Resistors in Combination

32 Chapter 18 Electric Charges in Motion

18.65

SET UP

A heater with a power rating of $P = 1500$ W is connected to a voltage source ($\varepsilon = 120$ V). The current drawn by the heater can be found using $P = iV$.

SOLVE

$$P = iV$$

$$i = \frac{P}{\varepsilon} = \frac{1500 \text{ W}}{120 \text{ V}} = \boxed{12 \text{ A (rounded to one significant figure)}}$$

REFLECT

Common household voltage in the United States is 120 V. Power ratings are calculated assuming this voltage.

Get Help: P'Cast 18.10 – Power in Series and Parallel Circuits

18.71

SET UP

The total energy required to run a 1500 W heater for 8 hours is equal to the product of the power and the time interval. Since we are given the cost the power company charges you in kWh ($0.11/kWh), we can multiply the numbers directly and then convert into kWh to find the total cost.

SOLVE

$$(1500 \text{ W})(8 \text{ h}) \times \frac{1 \text{ kWh}}{1000 \text{ W} \cdot \text{h}} \times \frac{\$0.11}{1 \text{ kWh}} = \boxed{\$1.32}$$

REFLECT

A kilowatt-hour is a unit of energy commonly used in electricity bills.

Get Help: P'Cast 18.10 – Power in Series and Parallel Circuits

18.73

SET UP

A resistor $R = 4.00 \times 10^6$ Ω and a capacitor $C = 3.00 \times 10^{-6}$ F are wired in series with a power supply. The time constant for this circuit is equal to the product RC.

SOLVE

$$RC = (4.00 \times 10^6 \text{ Ω})(3.00 \times 10^{-6} \text{ F}) = \boxed{12.0 \text{ s}}$$

REFLECT

The prefix *mega-* corresponds to 10^6 and *micro-* corresponds to 10^{-6}, so if we multiply them together, the powers of 10 will cancel.

Get Help: Interactive Example – RC I
Interactive Example – RC II
Interactive Example – RC III

Chapter 19
Magnetism

Conceptual Questions

19.1 Use both ends of one iron rod to approach the other iron rods. If both ends of the rod you are holding attract both ends of the other two rods, then the one you are holding is not magnetized iron.

19.5 Part a) The electric field points into the page.

Part b) The beam is deflected into the page.

Part c) The electron beam is not deflected.

Get Help: P'Cast 19.2 – Measuring Isotopes with a Mass Spectrometer

19.9 Since the directions of the two currents are opposite, the magnetic field from one wire is opposite to the other. This means the magnetic fields will cancel each other to some extent.

Get Help: P'Cast 19.3 – Magnetic Levitation

Multiple-Choice Questions

19.13 C (deflected toward the top of the page). The right-hand rule gives the direction of the force acting on a moving charge in a magnetic field.

Get Help: Interactive Example – Wall
Interactive Example – Fields
P'Cast 19.1 – Magnetic Forces on a Proton and an Electron

19.17 A (zero). The force on each infinitesimal portion of the ring points radially, so the torque on the loop is equal to zero.

19.21 E (zero force). You can prove this to yourself either through the right-hand rule or through symmetry.

Get Help: P'Cast 19.6 – Wires in a Computer

Estimation/Numerical Analysis

19.25 The time is equal to the distance traveled divided by the speed:

$$t = \frac{2\pi r}{v} = \frac{2\pi(0.1 \text{ m})}{\left(10^5 \frac{\text{m}}{\text{s}}\right)} \approx 6 \times 10^{-6} \text{ s} = 6 \text{ } \mu\text{s}$$

Get Help: Interactive Example – Wall
Interactive Example – Fields
P'Cast 19.1 – Magnetic Forces on a Proton and an Electron

Problems

19.29

SET UP

We are shown six scenarios of a positive charge moving with a velocity $\vec{v}$ in a magnetic field $\vec{B}$. We can use the right-hand rule to determine the direction of the magnetic force acting on the charge.

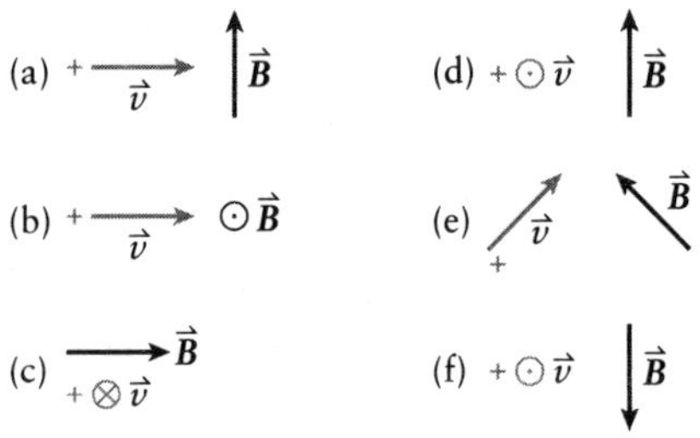

Figure 19-1 Problem 29

SOLVE

Part a) The force points out of the page.

Part b) The force points down.

Part c) The force points down.

Part d) The force points to the left.

Part e) The force points out of the page.

Part f) The force points to the right.

REFLECT

All of the answers would be reversed if the moving charge were negative.

Get Help: Interactive Example – Wall
Interactive Example – Fields
P'Cast 19.1 – Magnetic Forces on a Proton and an Electron

19.33

SET UP

A proton ($q = 1.60 \times 10^{-19}$ C) travels with a speed of 18 m/s toward the top of the page through a uniform magnetic field of 2.0 T directed into the page. The magnitude of the magnetic force acting on the proton is given by $F = |q|vB\sin\theta$, and the direction is given by the right-hand rule.

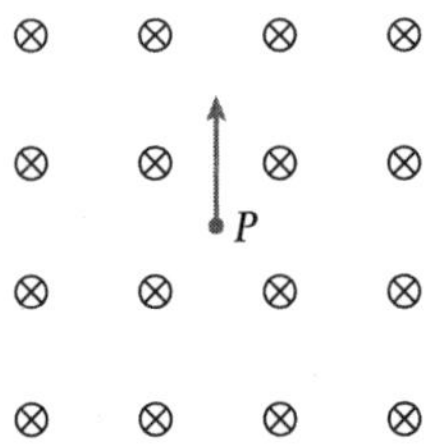

Figure 19-2 Problem 33

SOLVE

Magnitude:

$$F = |q|vB\sin\theta = (1.60 \times 10^{-19} \text{ C})\left(18\frac{\text{m}}{\text{s}}\right)(2.0 \text{ T})\sin(90°) = \boxed{5.8 \times 10^{-18} \text{ N}}$$

Direction: The force points to the $\boxed{\text{left}}$.

REFLECT

This is the maximum value of the force since $\vec{v}$ and $\vec{B}$ are perpendicular to one another.

Get Help: Interactive Example – Wall
Interactive Example – Fields
P'Cast 19.1 – Magnetic Forces on a Proton and an Electron

19.35

SET UP

An electron ($q = -1.60 \times 10^{-19}$ C) travels with a speed of 10^7 m/s in the x-y plane at an angle of 45° above both the +x and +y axes through a uniform magnetic field of 3.0 T in the +y direction. The magnitude of the force acting on the proton is given by the magnetic force law, $F = |q|vB\sin\theta$, and the direction of the force is given by the right-hand rule.

SOLVE
Magnitude:

$$F = |q|vB\sin\theta = (1.60 \times 10^{-19} \text{ C})\left(10^7 \frac{\text{m}}{\text{s}}\right)(3.0 \text{ T})\sin(45°) = \boxed{3.4 \times 10^{-12} \text{ N}}$$

Direction: The force points in the $\boxed{-z \text{ direction}}$.

REFLECT
Remember to take the sign of the charge into account when assigning the direction of the magnetic force.

Get Help: Interactive Example – Wall
Interactive Example – Fields
P'Cast 19.1 – Magnetic Forces on a Proton and an Electron

19.41

SET UP

A straight segment of wire that is 0.350 m long carries a current of 1.40 A in a uniform magnetic field. The segment makes an angle of 53° with $\vec{B}$. The magnitude of the force acting on the segment is 0.200 N. We can calculate the magnitude of the magnetic field from the expression for the magnetic force on a current-carrying wire, $F = i\ell B\sin\theta$.

SOLVE

$$F = i\ell B\sin\theta$$

$$B = \frac{F}{i\ell\sin\theta} = \frac{0.200 \text{ N}}{(1.40 \text{ A})(0.350 \text{ m})\sin(53°)} = \boxed{0.511 \text{ T}}$$

REFLECT
The force acts in a direction perpendicular to both the current flow and the magnetic field.

Get Help: P'Cast 19.3 – Magnetic Levitation

19.47

SET UP

A round loop of wire ($r = 0.10$ m) carries a current of $i = 100$ A. The loop makes an angle of $\phi = 30°$ with a magnetic field of $B = 0.244$ T. The angle is then changed to $\phi = 10°$ and $\phi = 50°$. The magnitude of the torque on the loop in each case is $\tau = iAB\sin\phi$.

SOLVE
$\phi = 30°$:

$$\tau = iAB\sin\phi = (100 \text{ A})(\pi(0.10 \text{ m})^2)(0.244 \text{ T})\sin(30°) = \boxed{0.4 \text{ N}\cdot\text{m}}$$

$\phi = 10°$:

$$\tau = iAB\sin\phi = (100\text{ A})(\pi(0.10\text{ m})^2)(0.244\text{ T})\sin(10°) = \boxed{0.1\text{ N} \cdot \text{m}}$$

$\phi = 50°$:

$$\tau = iAB\sin\phi = (100\text{ A})(\pi(0.10\text{ m})^2)(0.244\text{ T})\sin(50°) = \boxed{0.6\text{ N} \cdot \text{m}}$$

REFLECT

The torque is a maximum when $\phi = 90°$.

19.51

SET UP

A wire carrying a current i is bent into two semicircular parts and two flat parts (see figure). The magnetic field at point C is the vector sum of the magnetic fields due to each of the four segments. Since the flat parts lie along the same axis as C, they do not contribute to the field at C. We can use the results described in Section 19-7 of the text for the magnitude of the magnetic field along the axis of a current loop to determine the magnitude of the magnetic field produced by each semicircular wire. Because the wires are semicircular, we expect the magnitude of the magnetic fields to be one-half that of a closed circular loop. The magnetic field at point C due to the closer, smaller semicircle of radius r points into the page (which we'll call $-y$), whereas the magnetic field at point C due to the farther, larger semicircle of radius R points out of the page (or $+y$).

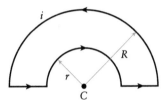

Figure 19-3 Problem 51

SOLVE

According to equation 19-18, the magnitude of the magnetic field along the axis of a current loop is given by:

$$B = \frac{\mu_0 i}{2} \frac{R^2}{(R^2 + y^2)^{3/2}}$$

where R is the radius of the loop and y is the distance from the center of the loop along the y axis. Since the point C is at the center of the circular loops, $y = 0$. Also, since the loops in this problem are semicircular, the field will be one-half of this quantity.

Magnitude of magnetic field due to a semicircular wire at its center:

$$B = \frac{1}{2}\left[\frac{\mu_0 i}{2} \frac{R^2}{(R^2 + (0)^2)^{3/2}}\right] = \frac{\mu_0 i}{4}\left(\frac{R^2}{R^3}\right) = \frac{\mu_0 i}{4R}$$

By the right-hand rule, we know that the magnetic field due to the inner arc, B_{inner}, points in the $-y$ direction and the field due to the outer arc, B_{outer}, points in the $+y$ direction. The radius of the inner loop is r and the radius of the outer loop is R. The magnitude of the field along the y axis, then, is:

$$B = -B_{\text{inner}} + B_{\text{outer}} = -\left(\frac{\mu_0 i}{4r}\right) + \left(\frac{\mu_0 i}{4R}\right) = \frac{\mu_0 i}{4}\left(\frac{1}{R} - \frac{1}{r}\right) = \frac{\mu_0 i}{4}\left(\frac{r - R}{rR}\right) = \boxed{-\frac{\mu_0 i}{4}\left(\frac{R - r}{rR}\right)}$$

REFLECT
The magnetic field at point C points into the page because $R > r$.

19.53

SET UP
A current of 100 A passes through a wire that is 5 m from a window. We can draw an Amperian loop with a radius $r = 5$ m around the wire and apply Ampère's law to find the magnitude of the magnetic field at the window. Once we know the magnitude of the field due to the power line, we can compare it to the magnitude of Earth's magnetic field ($B_{Earth} = 5 \times 10^{-5}$ T).

SOLVE

$$\sum B_{\parallel}\Delta \ell = \mu_0 i_{through}$$

$$B_{\parallel}\sum \Delta \ell = B(2\pi r) = \mu_0 i$$

$$B = \frac{\mu_0 i}{2\pi r} = \frac{\left(4\pi \times 10^{-7}\frac{T \cdot m}{A}\right)(100\ A)}{2\pi(5\ m)} = \boxed{4 \times 10^{-6}\ T}$$

$$\frac{B}{B_{Earth}} = \frac{4 \times 10^{-6}\ T}{5 \times 10^{-5}\ T} = 0.08$$

$$\boxed{B = 0.08 B_{Earth}}$$

REFLECT
Learning the algebraic expression for the magnitude of a magnetic field due to a straight, current-carrying wire will prove handy.

19.57

SET UP
Wire 1 ($\ell_1 = 0.0100$ m) carries a current of 2.00 A pointing north. Wire 2 carries a current of 3.60 A pointing south; wire 2 is $d = 1.40$ m to the right of wire 1. The magnitude of the force due to wire 2 on wire 1 is equal to $i_1 \ell_1 B_2 \sin\theta$. Because the wires are parallel, the angle θ is equal to 90°. As a reminder, the magnitude of a magnetic field a distance d away from a straight current-carrying wire is $\frac{\mu_0 i}{2\pi d}$. The direction of the force on wire 1 due to wire 2 is given by the right-hand rule between the direction of i_1 and the direction of B_2. The magnetic field due to wire 2 points directly into the page at the location of wire 1, so the force will point to the left.

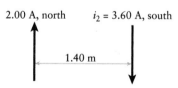

Figure 19-4 Problem 57

SOLVE
Magnitude:

$$F_{2\ on\ 1} = i_1 \ell_1 B_2 \sin\theta = i_1 \ell_1 B_2 \sin 90° = i_1 \ell_1 B_2 = i_1 \ell_1 \left(\frac{\mu_0 i_2}{2\pi d}\right) = \frac{\mu_0 i_1 i_2 \ell_1}{2\pi d}$$

$$= \frac{\left(4\pi \times 10^{-7}\frac{\text{T} \cdot \text{m}}{\text{A}}\right)(2.00\text{ A})(3.60\text{ A})(0.0100\text{ m})}{2\pi(1.40\text{ m})} = 1.03 \times 10^{-8}\text{ N}$$

The force on wire 1 due to wire 2 has a magnitude of $\boxed{1.03 \times 10^{-8}\text{ N and points to the left}}$.

REFLECT

In general, wires with antiparallel currents repel one another, while wires with parallel currents attract one another. You can prove this to yourself either by performing the similar calculations or by using Newton's third law.

Get Help: P'Cast 19.6 – Wires in a Computer

19.63

SET UP

A negatively charged particle enters a region of crossed electric and magnetic fields known as a velocity selector. The electric field has a magnitude of $E = 90{,}000$ V/m and points down (toward $-y$), and the magnetic field has a magnitude of $B_0 = 0.0053$ T and points into the page. In the velocity selector, the net force on the particle in the y direction is zero since the particle travels in a straight line. Therefore, the magnitude of the force due to the electric field must equal the magnitude of the force due to the magnetic field. Setting these equal gives us an expression for the speed of the particle as it exits the selector. Immediately after the selector, the particle enters a region with a magnetic field $B = 0.00242$ T. The particle then undergoes uniform circular motion in a radius $r = 0.0400$ m due to the magnetic force acting on it. We can solve for the mass of the particle by applying Newton's second law once again.

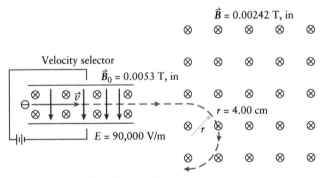

Figure 19-5 Problem 63

SOLVE

Free-body diagram of the particle in the velocity selector:

Figure 19-6 Problem 63

Speed of the particle exiting the velocity selector:

$$\sum F_{\text{ext},y} = F_{\text{magnetic},y} - F_{\text{electric},y} = ma_y = 0$$

$$qE - qvB_0 = 0$$

$$v = \frac{E}{B_0}$$

Free-body diagram of the particle as soon as it enters the spectrometer:

Figure 19-7 Problem 63

Mass of the particle:

$$\sum F_{ext,y} = -F_{magnetic} = ma_{cent} = m\left(-\frac{v^2}{r}\right)$$

$$|q|vB = \frac{mv^2}{r}$$

$$m = \frac{r|q|B}{v} = \frac{r|q|B_0 B}{E}$$

$$= \frac{(0.0400 \text{ m})(1.60 \times 10^{-19} \text{ C})(0.00242 \text{ T})(0.0053 \text{ T})}{\left(90{,}000\frac{\text{V}}{\text{m}}\right)} = \boxed{9 \times 10^{-31} \text{ kg}}$$

REFLECT

The accepted mass of the electron (to three significant figures) is 9.11×10^{-31} kg, so our value is reasonable.

Get Help: P'Cast 19.2 – Measuring Isotopes with a Mass Spectrometer

19.73

SET UP

Two straight conducting rods—rod 1 has a resistance $R_1 = 0.5 \, \Omega$, rod 2 has a resistance $R_2 = 2.5 \, \Omega$—that have the same mass $m = 25 \times 10^{-3}$ kg and length $\ell = 1.0$ m are connected in series by a resistor ($R = 17 \, \Omega$) and an external voltage source with potential difference ε_0. We are told that rod 1 "floats" a distance $d = 0.85 \times 10^{-3}$ m above rod 2, which means the net force acting on rod 1 is zero. The forces acting on rod 1 are the force due to the magnetic field generated by rod 2 pointing up and the force due to gravity pointing down. In order to find the magnetic field generated by rod 2, we first need to find the current in the circuit. The two rods and the resistor are in series, which means the current is constant throughout the circuit. The current is equal to the potential difference ε_0 divided by the equivalent resistance of the circuit. The magnitude of the magnetic field due to rod 2 at rod 1 is $B_2 = \frac{\mu_0 i}{2\pi d}$, and the magnitude of the magnetic force acting on rod 2 is $i\ell B_2$. Setting this expression equal to the force of gravity, we can calculate ε_0.

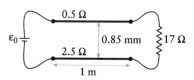

Figure 19-8 Problem 73

SOLVE
Equivalent resistance:
$$R_{equiv} = R_1 + R_2 + R = (0.5\ \Omega) + (2.5\ \Omega) + (17\ \Omega) = 20.0\ \Omega$$

Current in the circuit:
$$V = iR$$
$$\varepsilon_0 = iR_{equiv}$$
$$i = \frac{\varepsilon_0}{R_{equiv}}$$

Magnetic field due to rod 2 at the location of rod 1:
$$B_2 = \frac{\mu_0 i}{2\pi d} = \frac{\mu_0 \left(\frac{\varepsilon_0}{R_{equiv}}\right)}{2\pi d} = \frac{\mu_0 \varepsilon_0}{2\pi d R_{equiv}}$$

Free-body diagram for rod 2:

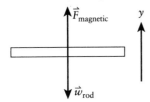

Figure 19-9 Problem 73

Newton's second law for rod 1:
$$\sum F_{ext,y} = F_{magnetic} - w_{rod} = m_{rod} a_y = 0$$
$$F_{magnetic} = w_{rod}$$
$$i\ell B_2 = m_{rod} g$$
$$\left(\frac{\varepsilon_0}{R_{equiv}}\right)\ell\left(\frac{\mu_0 \varepsilon_0}{2\pi d R_{equiv}}\right) = m_{rod} g$$

$$\varepsilon_0 = \sqrt{\frac{2\pi d R_{equiv}^2 m_{rod} g}{\mu_0 \ell}} = \sqrt{\frac{2\pi (0.85 \times 10^{-3}\ \text{m})(20.0\ \Omega)^2 (25 \times 10^{-3}\ \text{kg})\left(9.80 \frac{\text{m}}{\text{s}^2}\right)}{\left(4\pi \times 10^{-7} \frac{\text{T}\cdot\text{m}}{\text{A}}\right)(1.0\ \text{m})}} = \boxed{650\ \text{V}}$$

REFLECT
In the space directly above rod 2, its magnetic field points into the page. The current through rod 1 is moving toward the right.

Get Help: P'Cast 19.6 – Wires in a Computer

19.77

SET UP

A high-voltage power line is located 5.0 m in the air and 12 m horizontally from your house. The wire carries a current of 100 A. We can use the expression $B = \dfrac{\mu_0 i}{2\pi r}$, where r is the straight-line distance from the wire to your house, to calculate the magnitude of the magnetic field caused by the power line. Once we know the magnitude of the field due to the power line, we can compare it to the magnitude of Earth's magnetic field ($B_{\text{Earth}} = 0.5 \times 10^{-4}$ T) to determine if the power line is likely to affect your health.

SOLVE

$$B = \dfrac{\mu_0 i}{2\pi r} = \dfrac{\left(4\pi \times 10^{-7} \dfrac{\text{T} \cdot \text{m}}{\text{A}}\right)(100 \text{ A})}{2\pi \sqrt{(5.0)^2 + (12 \text{ m})^2}}$$

$$= 1.54 \times 10^{-6} \text{ T} \approx \boxed{2 \times 10^{-6} \text{ T (rounded to 1 significant figure)}}$$

$$\dfrac{B}{B_{\text{Earth}}} = \dfrac{1.54 \times 10^{-6} \text{ T}}{0.5 \times 10^{-4} \text{ T}} = 0.03$$

$$\boxed{B = 0.03 B_{\text{Earth}}}$$

Since the magnetic field from the wires is so much smaller than Earth's magnetic field, there should be $\boxed{\text{little or no cause for concern}}$.

REFLECT

Even if we were 1 m from the power line, the magnitude of the magnetic field would still only be 40% of the strength of Earth's field.

Get Help: P'Cast 19.6 – Wires in a Computer

Chapter 20
Electromagnetic Induction

Conceptual Questions

20.1 The falling bar magnet induces a current in the wall of the pipe. In accordance with Lenz's law, the direction of this induced current is such that it produces a magnetic field that exerts a force on the magnet opposing its motion. The speed of the magnet therefore increases more slowly than in free-fall until it reaches a terminal speed at which the magnetic and gravitational forces are equal and opposite. After this point, the magnet continues to fall, but at a constant speed.

Multiple-Choice Questions

20.7 D (leaving the magnetic field). The loop is starting to exit the region with the magnetic field.

(d)

Figure 20-1 Problem 7

> **Get Help:** Interactive Example – Motion emf
> P'Cast 20.2 – Changing Magnetic Flux II: A Varying Magnetic Field

20.9 C (antiparallel to i_a). The induced current in loop b will flow in such a way as to generate a magnetic field to counteract the increase in flux due to loop a.

> **Get Help:** Interactive Example – Motion emf
> P'Cast 20.2 – Changing Magnetic Flux II: A Varying Magnetic Field

Problems

20.23

SET UP

A bar magnet, leading with its south pole, is moved at a constant velocity through a wire loop. We'll assume the magnet starts infinitely far to the left of the loop at $t = 0$. As the bar magnet passes through the loop, the magnetic flux is constantly changing, which induces a voltage in the loop according to Faraday's law and Lenz's law. In order to draw a qualitative sketch of the induced voltage in the loop, we will

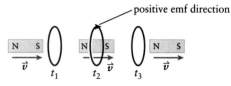

Figure 20-2 Problem 23

define the area vector of the loop to point toward the initial location of the magnet (to the left in the above figure); this defines our sense of positive induced voltage. The magnetic field pointing to the left is getting stronger as the magnet nears the loop before t_1. Since this corresponds to an increasing positive flux, this will generate an increasingly negative induced voltage until it reaches a minimum. Similarly, after the magnet leaves the loop, the magnetic field pointing to the left is getting weaker. This corresponds to a decreasing positive flux, which results in an increasingly positive induced voltage until it reaches a maximum. The flux is changing the most at t_2, so this corresponds to a maximum in the plot of the magnetic flux versus time. A maximum in the flux corresponds to an induced voltage of zero since the induced current will change direction at that moment.

SOLVE

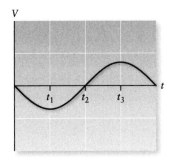

Figure 20-3 Problem 23

REFLECT
The induced voltage is related to the slope of the magnetic flux as a function of time.

20.27

SET UP
An electromagnetic generator consists of a wound coil ($N = 100$) that has an area of $A = 400$ cm². The coil rotates at $60\frac{\text{rev}}{\text{s}}$ in a magnetic field with strength $B = 0.25$ T. The maximum induced potential is given by $\varepsilon_{max} = \omega NAB$, where ω is in rad/s.

SOLVE
$$\varepsilon = \omega NAB \sin(\omega t)$$
$$\varepsilon_{max} = \omega NAB$$
$$= \left(60\frac{\text{rev}}{\text{s}}\right)\left(\frac{2\pi \text{ rad}}{\text{rev}}\right)(100)\left(400 \text{ cm}^2 \times \left(\frac{1 \text{ m}}{100 \text{ cm}}\right)^2\right)(0.25 \text{ T})$$
$$= 4 \times 10^2 \text{ V}$$

The magnitude of the induced potential is 400 V.

REFLECT
This is the maximum value; the actual potential oscillates between $+400$ V and -400 V.

Get Help: Interactive Example – Single Turn
P'Cast 20.3 – Lighting the Gym with a Bicycle

20.31

SET UP

A magnetic field of $B = 0.45 \times 10^{-4}$ T is directed straight down, perpendicular to the plane of a circular coil of wire. The wire is made up of $N = 250$ turns and has an initial radius of $r_1 = 0.2$ m. The radius of the circle is increased to $r_2 = 0.3$ m in a period of $\Delta t = 15 \times 10^{-3}$ s. For simplicity, we'll assume the rate at which the radius changes is constant, the number of coils remains constant, and that the resistance of the coil also remains constant at $R = 25 \, \Omega$ throughout the stretch. We can use Faraday's law of induction, $|\varepsilon| = \left|\dfrac{\Delta \Phi_B}{\Delta t}\right|$, to calculate the voltage induced across the coil. Once we have the induced voltage across the coil, we can use Ohm's law to calculate the inducted current in the coil. Finally, the flux directed downward through the coil is increasing when the area of the loop is increasing; the induced magnetic field should point upward, so as to counteract this increase in flux. Therefore, the induced current will flow counterclockwise (when viewed from above), as given by the right-hand rule.

SOLVE

Part a)

$$|\varepsilon| = \left|\dfrac{\Delta \Phi_B}{\Delta t}\right| = \dfrac{\Delta(NAB \cos\theta)}{\Delta t} = \dfrac{\Delta(NAB \cos 0)}{\Delta t}$$

$$= NB\dfrac{\Delta A}{\Delta t} = NB\dfrac{(\pi r_f^2 - \pi r_i^2)}{\Delta t} = NB\pi\dfrac{(r_f^2 - r_i^2)}{\Delta t}$$

$$= (250)(0.45 \times 10^{-4} \text{ T})\pi \left(\dfrac{(0.3 \text{ m})^2 - (0.2 \text{ m})^2}{15 \times 10^{-3} \text{ s}}\right)$$

$$= \boxed{0.118 \text{ V} \approx 0.1 \text{ V (rounded to one significant figure)}}$$

Part b)

$$\varepsilon = iR$$

$$i = \dfrac{\varepsilon}{R} = \dfrac{0.118 \text{ V}}{25 \, \Omega} = \boxed{0.005 \text{ A}}$$

Part c) The induced current in the loop is $\boxed{\text{counterclockwise}}$ when viewed from above.

REFLECT

Our assumption that the resistance remained constant made the problem much easier to solve, but was an oversimplification. The resistance of the wire would increase linearly as we increase the radius of the coil because we've changed the dimensions of the coil.

Get Help: Interactive Example – Single Turn
P'Cast 20.3 – Lighting the Gym with a Bicycle

20.35

SET UP

A pair of parallel conducting rails are a distance $L = 0.12$ m apart and situated at right angles to a uniform magnetic field $B = 0.8$ T pointed into the page. A resistor ($R = 15$ Ω) is connected across the rails. A conducting bar is placed on top of the rails and is moved at a constant speed of $v = 2$ m/s to the right. The bar creates a closed loop with a portion of each rail and the resistor. The area of the loop is equal to Lx, where x is the horizontal distance between the resistor and the bar; we'll assume the area vector also points into the page. The area of this closed loop increases as the bar moves to the right, which means the magnetic flux is also increasing. We can use Faraday's law and Ohm's law to calculate the magnitude of the current flowing through the resistor. Since the magnetic flux into the page is increasing, current will be induced in such a way as to counteract this increase according to Lenz's law. From the right-hand rule, this corresponds to a counterclockwise current through the loop, so the current through the bar points up. The bar will experience a magnetic force due to the external magnetic field interacting with the current. The current is perpendicular to the field. In this case, the magnitude of the force is given by $F = iLB$, and the direction is given by the right-hand rule.

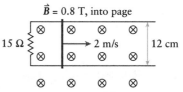

Figure 20-4 Problem 35

SOLVE

Part a)

Induced potential:

$$\varepsilon = \left|\frac{\Delta \Phi_B}{\Delta t}\right| = \left|\frac{\Delta(BA\cos\theta)}{\Delta t}\right| = \left|\frac{\Delta(BA\cos 0)}{\Delta t}\right| = \left|B\frac{\Delta A}{\Delta t}\right| = \left|B\frac{\Delta(Lx)}{\Delta t}\right| = \left|BL\frac{\Delta x}{\Delta t}\right|$$

$$\varepsilon = |BLv| = (0.8 \text{ T})(0.12 \text{ m})\left(2\frac{\text{m}}{\text{s}}\right) = 0.192 \text{ V}$$

Current:

$$\varepsilon = iR$$

$$i = \frac{\varepsilon}{R} = \frac{0.192 \text{ V}}{15 \text{ Ω}} = \boxed{0.0128 \text{ A} \approx 0.01 \text{ A (rounded to one significant figure)}}$$

Part b) The current in the bar points $\boxed{\text{up}}$.

Part c)

$$F = iLB = (0.0128 \text{ A})(0.12 \text{ m})(0.8 \text{ T}) = 0.001 \text{ N}$$

The magnetic force on the bar has a magnitude of $\boxed{0.001 \text{ N and points to the left}}$.

REFLECT

By defining our area vector to point down, we've implicitly chosen a clockwise current to be positive. The negative sign in our induced potential means we have a "negative" current in the loop, *i.e.*, one that flows counterclockwise.

Get Help: Interactive Example – Single Turn
P'Cast 20.3 – Lighting the Gym with a Bicycle

Chapter 21
Alternating-Current Circuits

Conceptual Questions

21.1 It means that the current peaks one-fourth of a period after the voltage drop peaks. One-fourth of a period corresponds to a phase difference of 1/4 cycle.

Get Help: P'Cast 21.3 – How Large Is One Henry?
P'Cast 21.4 – Inductor Voltage

21.11 The electrical energy stored in the capacitor is $U_E = \dfrac{q^2}{2C}$, which varies as $\cos^2(\omega t)$, and the magnetic energy stored in the inductor is $U_B = \dfrac{1}{2}Li^2$, which varies as $\sin^2(\omega t)$.

Therefore, when a maximum amount of energy is stored in the capacitor, no energy is stored in the inductor, and *vice versa*.

21.15 Part a) Assuming the number of coils remains constant, the self-inductance drops by a factor of 2.

$$L = \frac{N\Phi_B}{i} = \frac{N}{i}\left(\frac{\mu_0 N i}{\ell}\right)A = \frac{\mu_0 N^2 A}{\ell} \sim \frac{1}{\ell}$$

$$\frac{L_2}{L_1} = \frac{\frac{1}{(2\ell)}}{\frac{1}{\ell}} = \frac{1}{2}$$

Part b) Assuming the length of the solenoid remains constant, the self-inductance is unchanged.

$$L = \frac{\mu_0 N^2 A}{\ell_{solenoid}} = \frac{\mu_0}{\ell_{solenoid}}\left(\frac{\ell_{wire}}{\pi D_{solenoid}}\right)^2 \pi D^2_{solenoid} = \frac{\mu_0 \ell^2_{wire}}{\pi \ell_{solenoid}}$$

Get Help: P'Cast 21.3 – How Large Is One Henry?
P'Cast 21.4 – Inductor Voltage

Multiple-Choice Questions

21.19 D ($2\sqrt{2}V$). The peak-to-peak voltage is twice the peak voltage. The peak voltage is $\sqrt{2}$ times the root mean square voltage.

Get Help: P'Cast 21.1 – Traveling Abroad with Your Hair Dryer

21.23 C (1/4 of the total energy).

$$U_E = \frac{q_{max}^2}{2C}$$

$$\frac{U_{E,\text{half max}}}{U_{E,\text{max}}} = \frac{\left(\frac{1}{2C}\left(\frac{q_{max}}{2}\right)^2\right)}{\left(\frac{1}{2C}q_{max}^2\right)} = \frac{1}{4}$$

Estimation/Numerical Analysis

21.27 Most household appliances draw between 1 and 20 amps. Most households have a main circuit breaker that "trips" at 150 A or so.

Get Help: P'Cast 21.1 – Traveling Abroad with Your Hair Dryer

Problems

21.33

SET UP

The peak voltage across the terminals of a sinusoidal ac source ($f = 60.0$ Hz) is $V_0 = +17.0$ V. The voltage at $t = 0$ is equal to 0, which means we can model the voltage as a function of time by $V(t) = V_0 \sin(\omega t) = V_0 \sin((2\pi f)t)$. Using the provided information and the algebraic form for $V(t)$, we can solve for the voltage at $t = 2.00 \times 10^{-3}$ s.

SOLVE

$$V(t) = V_0 \sin\omega t$$

$$V(t = 2.00 \times 10^{-3}\text{ s}) = (17.0\text{ V})\sin(2\pi(60.0\text{ Hz})(2.00 \times 10^{-3}\text{ s})) = \boxed{11.6\text{ V}}$$

REFLECT

The maximum voltage occurs at 1/4 cycle or $t = 0.00417$ s, which means $V(t = 0.002\text{ s})$ should be positive and less than 17 V.

Get Help: P'Cast 21.1 – Traveling Abroad with Your Hair Dryer

21.39

SET UP

The primary coil ($N_p = 400$ turns) of a step-down transformer is connected to an ac line that has an applied voltage of $V_p = 120$ V. The secondary coil voltage is $V_s = 6.50$ V. The relationship between the number of turns and the voltages of each coil is $V_s = \frac{N_s}{N_p}V_p$. We can rearrange this equation to calculate the number of turns of the secondary coil N_s.

SOLVE

$$\frac{V_s}{V_p} = \frac{N_s}{N_p}$$

48 Chapter 21 Alternating-Current Circuits

$$N_s = \frac{V_s N_p}{V_p} = \frac{(6.50 \text{ V})(400 \text{ turns})}{120 \text{ V}} = 21.7 = \boxed{22 \text{ turns}}$$

REFLECT

We are told this is a step-down transformer, which means the $N_s < N_p$.

Get Help: Interactive Example – Square Loop
P'Cast 21.2 – A High-Voltage Transformer

21.43

SET UP

The primary coil ($N_p = 400$ turns) of a step-down transformer is connected to an ac line that has an applied voltage of $V_p = 120$ V. The secondary coil voltage is $V_s = 6.30$ V. The relationship between the number of turns and the voltages of each coil is $\frac{V_s}{V_p} = \frac{N_s}{N_p}$. We can rearrange this equation to calculate the number of turns of the secondary coil N_s. The secondary coil supplies a current of $i_s = 15.0$ A. We can use the relationship between the currents and the voltages of the two coils, $i_p V_p = i_s V_s$, to calculate the current in the primary coil.

SOLVE

Part a)

$$\frac{V_s}{V_p} = \frac{N_s}{N_p}$$

$$N_s = \frac{V_s N_p}{V_p} = \frac{(6.30 \text{ V})(400 \text{ turns})}{120 \text{ V}} = \boxed{21 \text{ turns}}$$

Part b)

$$i_p V_p = i_s V_s$$

$$i_p = \frac{V_s}{V_p} i_s = \left(\frac{6.30 \text{ V}}{120 \text{ V}}\right)(15.0 \text{ A}) = \boxed{0.79 \text{ A}}$$

REFLECT

A step-down transformer is designed such that the primary coil has a high voltage but low current and the secondary coil has a low voltage but high current.

Get Help: Interactive Example – Square Loop
P'Cast 21.2 – A High-Voltage Transformer

21.47

SET UP

A radio antenna is made from a solenoid of length $l = 0.030$ m and cross-sectional area $A = 0.50$ cm². The solenoid, which is filled with air, consists of $N = 300$ turns of copper wire. The inductance of a solenoid is $L = \frac{\mu_0 N^2 A}{\ell}$.

SOLVE

$$L = \frac{\mu_0 N^2 A}{\ell} = \frac{\left(4\pi \times 10^{-7} \frac{\text{H}}{\text{m}}\right)(300)^2 \left(0.50 \text{ cm}^2 \times \left(\frac{1 \text{ m}}{100 \text{ cm}}\right)^2\right)}{(0.030 \text{ m})}$$

$$= \boxed{1.9 \times 10^{-4} \text{ H} = 0.19 \text{ mH}}$$

REFLECT

The inductance of a solenoid only depends on its dimensions and a physical constant, so the fact that the wire is copper does not affect the inductance.

Get Help: P'Cast 21.3 – How Large Is One Henry?
P'Cast 21.4 – Inductor Voltage

21.57

SET UP

A circuit that contains a resistor, an inductor ($L = 5.00$ H), and a capacitor resonates at $f_0 = 1000$ Hz. Using the expression for the natural frequency of an LRC circuit, $\omega_0 = \sqrt{\frac{1}{LC}}$, we can calculate the value of the capacitance. Recall that $\omega_0 = 2\pi f_0$.

SOLVE

$$\omega_0 = \sqrt{\frac{1}{LC}}$$

$$C = \frac{1}{L\omega_0^2} = \frac{1}{L(2\pi f_0)^2} = \frac{1}{4\pi^2 (5.00 \text{ H})(1000 \text{ Hz})^2} = \boxed{5 \times 10^{-9} \text{ F} = 5 \text{ nF}}$$

REFLECT

The resistor does not affect the natural frequency of the circuit, only the maximum current.

Get Help: Interactive Example – AC 1
P'Cast 21.6 – Tuning an FM Radio

21.63

SET UP

A sinusoidal voltage ($V_{\text{rms}} = 40.0$ V, $f = 100$ Hz) is applied to a resistor ($R = 100$ Ω), an inductor ($L = 0.200$ H), and a capacitor ($C = 50.0 \times 10^{-6}$ F). The current as a function of time for a resistor, inductor, and capacitor each attached separately to an ac voltage source is $i_R(t) = \frac{V_0}{R}\sin(\omega t)$, $i_L(t) = \frac{V_0}{\omega L}\cos(\omega t)$, and $i_C(t) = \omega C V_0 \cos(\omega t)$, respectively. The peak current is the amplitude of the oscillation in all cases. The average power delivered to the resistor is equal to $P_{R,\text{average}} = \frac{1}{2}\frac{V_0^2}{R}$. Since the power is equal to the voltage multiplied by the current, the average power delivered to the inductor and capacitor is zero because the current and voltage are 1/4 cycle out of phase.

SOLVE

Part a)

Peak current:

$$i_R(t) = \frac{V_0}{R} \sin(\omega t)$$

$$i_{R,0} = \frac{V_0}{R} = \frac{V_{rms}\sqrt{2}}{R} = \frac{(40.0\ \text{V})\sqrt{2}}{100\ \Omega} = \boxed{0.6\ \text{A}}$$

Average power:

$$P_{R,average} = \frac{1}{2}\frac{V_0^2}{R} = \frac{1}{2}\frac{(V_{rms}\sqrt{2})^2}{R} = \frac{V_{rms}^2}{R} = \frac{(40.0\ \text{V})^2}{100\ \Omega}$$

$$= \boxed{16\ \text{W} = 2 \times 10^1\ \text{W (rounded to 1 significant figure)}}$$

Part b)

Peak current:

$$i_L(t) = \frac{V_0}{\omega L} \cos(\omega t)$$

$$i_{L,0} = \frac{V_0}{\omega L} = \frac{V_{rms}\sqrt{2}}{(2\pi f)L} = \frac{(40.0\ \text{V})\sqrt{2}}{2\pi(100\ \text{Hz})(0.200\ \text{H})} = \boxed{0.5\ \text{A}}$$

Average power:

$$P_{L,average} = 0$$

Part c)

Peak current:

$$i_C(t) = \omega C V_0 \cos(\omega t)$$

$$i_{C,0} = \omega C V_0 = (2\pi f)C(V_{rms}\sqrt{2}) = 2\pi\sqrt{2}fCV_{rms}$$

$$= 2\pi\sqrt{2}(100\ \text{Hz})(50.0 \times 10^{-6}\ \text{F})(40.0\ \text{V}) = \boxed{2\ \text{A}}$$

Average power:

$$P_{C,average} = 0$$

REFLECT

Power for both the inductor and the capacitor is alternately positive and negative because energy is alternately flowing into and out of them. The *average* power is therefore zero for both of them.

Get Help: Interactive Example – AC 1
P'Cast 21.6 – Tuning an FM Radio

21.65

SET UP

An ac voltage, described by $V(t) = (10 \text{ V}) \sin(12\pi t)$, is applied to an LRC series circuit with $L = 0.25$ H, $R = 20\,\Omega$, and $C = 350 \times 10^{-6}$ F. We want to know the voltage across each element at a time $t = 0.04$ s. From $V(t)$, we know that $V_0 = 10$ V and $\omega = 12\pi$. The voltage for each element is given by the equation $V(t) - L\dfrac{\Delta i(t)}{\Delta t} - i(t)R - \dfrac{q(t)}{C} = 0$, where $i(t) = \dfrac{V_0}{Z}\sin(\omega t + \phi)$. The impedance is $Z = \sqrt{\left(\dfrac{1}{\omega C} - \omega L\right)^2 + R^2}$, and the phase angle is $\phi = \arctan\left[\dfrac{1}{R}\left(\dfrac{1}{\omega C} - \omega L\right)\right]$. The charge and the rate of change of the current can be determined by inspection of the current (using the analogy of equations 12-6, 12-7, and 12-8): $q(t) = \dfrac{-V_0}{\omega Z}\cos(\omega t + \phi)$ and $\dfrac{\Delta i(t)}{\Delta t} = \dfrac{\omega V_0}{Z}\cos(\omega t + \phi)$.

SOLVE

Impedance:

$$Z = \sqrt{\left(\dfrac{1}{\omega C} - \omega L\right)^2 + R^2}$$

$$= \sqrt{\left(\dfrac{1}{\left(12\pi \dfrac{\text{rad}}{\text{s}}\right)(350 \times 10^{-6} \text{ F})} - \left(12\pi \dfrac{\text{rad}}{\text{s}}\right)(0.25 \text{ H})\right)^2 + (20\,\Omega)^2}$$

$$= 69.312\,\Omega$$

Phase angle:

$$\phi = \arctan\left[\dfrac{1}{R}\left(\dfrac{1}{\omega C} - \omega L\right)\right]$$

$$= \arctan\left[\dfrac{1}{(20\,\Omega)}\left(\dfrac{1}{\left(12\pi \dfrac{\text{rad}}{\text{s}}\right)(350 \times 10^{-6} \text{ F})} - \left(12\pi \dfrac{\text{rad}}{\text{s}}\right)(0.25 \text{ H})\right)\right]$$

$$= 1.278 \text{ rad}$$

Voltage:

$$V(t) - L\dfrac{\Delta i(t)}{\Delta t} - i(t)R - \dfrac{q(t)}{C} = 0$$

$$V_L(t) = L\dfrac{\Delta i(t)}{\Delta t} \qquad V_R(t) = i(t)R \qquad V_C(t) = \dfrac{q(t)}{C}$$

Chapter 21 Alternating-Current Circuits

Current, charge, and rate of change of current:

$$i(t) = \frac{V_0}{Z}\sin(\omega t + \phi)$$

$$q(t) = \frac{-V_0}{\omega Z}\cos(\omega t + \phi)$$

$$\frac{\Delta i(t)}{\Delta t} = \frac{\omega V_0}{Z}\cos(\omega t + \phi)$$

Voltage across the inductor at $t = 0.04$ s:

$$V_L(t) = L\frac{\Delta i(t)}{\Delta t} = \frac{L\omega V_0}{Z}\cos(\omega t + \phi)$$

$$V_L(0.04 \text{ s}) = \frac{(0.25 \text{ H})\left(12\pi\frac{\text{rad}}{\text{s}}\right)(10 \text{ V})}{(69.312 \text{ }\Omega)}\cos\left(\left(12\pi\frac{\text{rad}}{\text{s}}\right)(0.04 \text{ s}) + (1.278 \text{ rad})\right)$$

$$= \boxed{-1.27 \text{ V}}$$

Voltage across the resistor at $t = 0.04$ s:

$$V_R(t) = i(t)R = \frac{V_0 R}{Z}\sin(\omega t + \phi)$$

$$V_R(0.04 \text{ s}) = \frac{(10 \text{ V})(20 \text{ }\Omega)}{(69.312 \text{ }\Omega)}\sin\left(\left(12\pi\frac{\text{rad}}{\text{s}}\right)(0.04 \text{ s}) + (1.278 \text{ rad})\right)$$

$$= \boxed{1.00 \text{ V}}$$

Voltage across the capacitor at $t = 0.04$ s:

$$V_C(t) = \frac{q(t)}{C} = -\frac{V_0}{C\omega Z}\cos(\omega t + \phi)$$

$$V_C(0.04 \text{ s}) = -\frac{(10 \text{ V})}{(350 \times 10^{-6} \text{ F})\left(12\pi\frac{\text{rad}}{\text{s}}\right)(69.312 \text{ }\Omega)}\cos\left(\left(12\pi\frac{\text{rad}}{\text{s}}\right)(0.04 \text{ s}) + (1.278 \text{ rad})\right)$$

$$= \boxed{10.3 \text{ V}}$$

REFLECT

We can check our answer by adding the voltages across each element at $t = 0.04$ s:

$$V(0.04 \text{ s}) = V_L(0.04 \text{ s}) + V_R(0.04 \text{ s}) + V_C(0.04 \text{ s})$$

$$V(0.04 \text{ s}) = (-1.27 \text{ V}) + (1.00 \text{ V}) + (10.3 \text{ V}) \approx 10 \text{ V}$$

Since the sum of the voltages across each element at $t = 0.04$ s equals the total voltage at $t = 0.04$ s, our answer is reasonable.

Get Help: Interactive Example – AC 1
P'Cast 21.6 – Tuning an FM Radio

21.73

SET UP

A circuit contains a resistor ($R = 500\ \Omega$), an inductor ($L = 5.00$ H), and a capacitor wired in series. Using the expression for the natural frequency of an LRC circuit, $\omega_0 = \sqrt{\dfrac{1}{LC}}$, we can calculate the capacitance that will cause the circuit to resonate at a frequency of $f_0 = 1000$ Hz. Recall that $\omega_0 = 2\pi f_0$.

SOLVE

$$\omega_0 = 2\pi f_0 = \sqrt{\dfrac{1}{LC}}$$

$$C = \dfrac{1}{L(2\pi f_0)^2} = \dfrac{1}{4\pi^2 f_0^2 L} = \dfrac{1}{4\pi^2 (1000\ \text{Hz})^2 (5.00\ \text{H})} = \boxed{5 \times 10^{-9}\ \text{F}}$$

REFLECT

The resistor does not affect the natural frequency of the circuit, only the maximum current.

Get Help: Interactive Example – AC 1
P'Cast 21.6 – Tuning an FM Radio

21.79

SET UP

An ac generator is connected across a light bulb ($R = 8.50\ \Omega$). The electrical generator is made by rotating a flat coil in a uniform magnetic field ($B = 0.225$ T). The flat coil has 33 windings and is square, measuring 0.150 m on each side. The coil rotates at $\omega = 745\,\dfrac{\text{rev}}{\text{min}}$ about an axis perpendicular to the magnetic field and parallel to its two opposite sides. This orientation means the coil will rotate around at a frequency of ω. The alternating induced potential in the coil can be determined through Faraday's law for emf induced in an ac generator, which is $\varepsilon = \omega NAB \sin(\omega t + \phi)$, where ϕ is the angle between the magnetic field and the perpendicular to the coil. We'll assume this angle is initially zero, $\phi = 0$.
The amplitude of the resulting function is also equal to the voltage amplitude V_0 across the light bulb. The current amplitude for the bulb is equal to $i_0 = \dfrac{V_0}{R}$ through Ohm's law. The average rate heat is generated is equal to the average power emitted by the light bulb, $P_{\text{average}} = i_{\text{rms}}^2 R$, where $i_{\text{rms}} = \dfrac{i_0}{\sqrt{2}}$. The energy consumed by the light bulb in an hour is equal to its power multiplied by 1 hour.

SOLVE

Part a)

Angular frequency:

$$\omega = 745\frac{\text{rev}}{\text{min}} \times \frac{1\text{ min}}{60\text{ s}} \times \frac{2\pi\text{ rad}}{1\text{ rev}} = \frac{745\pi}{30}\frac{\text{rad}}{\text{s}}$$

Induced potential:

$$\varepsilon = \omega NAB\sin(\omega t + \phi)$$

$$= \left(\frac{745\pi}{30}\frac{\text{rad}}{\text{s}}\right)(33)(0.150\text{ m})^2(0.225\text{ T})\sin(\omega t + 0)$$

$$= (13.03\text{ V})\sin\omega t$$

$$V_0 = 13.03 \approx \boxed{13.0\text{ V}}$$

Current:

$$i_0 = \frac{V_0}{R} = \frac{13.03\text{ V}}{8.50\text{ }\Omega} = 1.533\text{ A} \approx \boxed{1.53\text{ A}}$$

Part b)

$$P_{\text{average}} = i_{\text{rms}}^2 R = \left(\frac{i_0}{\sqrt{2}}\right)^2 R = \frac{(1.533\text{ A})^2}{2}(8.50\text{ }\Omega) = 9.988\text{ W} \approx \boxed{9.99\text{ W}}$$

Part c)

$$P_{\text{average}} = \frac{\Delta E_{\text{consumed}}}{\Delta t}$$

$$\Delta E_{\text{consumed}} = P_{\text{average}}\Delta t = (9.988\text{ W})\left(1\text{ hr} \times \frac{3600\text{ s}}{1\text{ hr}}\right) = \boxed{3.60 \times 10^4\text{ J}}$$

REFLECT

In the United States, we buy light bulbs based on the power they consume at a voltage of 120 V rather than their resistance. Accordingly, a 60-W light bulb has a resistance of 240 Ω.

Get Help: P'Cast 21.1 – Traveling Abroad with Your Hair Dryer

21.83

SET UP

The circuit seen in the figure is known as a low-pass filter. The input is an ac signal composed of many frequencies. The output detected is the voltage across the capacitor. The current in the circuit is equal to the input voltage divided by the impedance of the circuit. The total impedance of a circuit with an inductor, a capacitor, and a resistor in series is $Z = \sqrt{\left(\frac{1}{\omega C} - \omega L\right)^2 + R^2}$. For a circuit with just a resistor and a capacitor

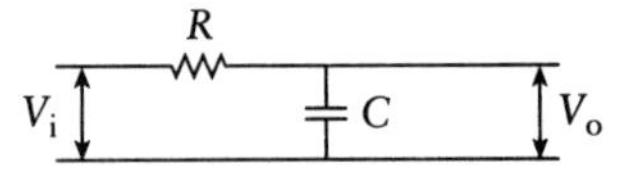

Figure 21-1 Problem 83

($L = 0$), the impedance is $Z = \sqrt{\left(\dfrac{1}{\omega C}\right)^2 + R^2}$. The impedance of the capacitor alone is $Z = \dfrac{1}{\omega C}$. We can then use the impedance of the capacitor as well as the expression for the voltage across the capacitor $V_o = i_0\left(\dfrac{1}{\omega C}\right)$ to solve for the ratio between the output and input voltages.

SOLVE
Current:

$$V_0 = i_0 Z$$

$$i_0 = \frac{V_i}{Z} = \frac{V_i}{\sqrt{\left(\dfrac{1}{\omega C}\right)^2 + R^2}}$$

Voltage across the capacitor:

$$V_o = i_0 Z_C = i_0\left(\frac{1}{\omega C}\right) = \left(\frac{V_i}{\sqrt{\left(\dfrac{1}{\omega C}\right)^2 + R^2}}\right)\left(\frac{1}{\omega C}\right)$$

Ratio between V_o and V_i:

$$\frac{V_o}{V_i} = \frac{1}{\omega C \sqrt{\left(\dfrac{1}{\omega C}\right)^2 + R^2}} = \frac{1}{\sqrt{(\omega C)^2\left(\dfrac{1}{\omega C}\right)^2 + (\omega C)^2(R^2)}} = \boxed{\frac{1}{\sqrt{1 + (\omega RC)^2}}}$$

REFLECT
Looking at our final expression for the ratio of the voltages, we see the $\dfrac{V_o}{V_i}$ gets small as f gets large, which means the amplitude of the high-frequency components of the input signal is very small. This is consistent with our expectation that a low-pass filter allows low-frequency signals to pass (relatively) unaffected while attenuating the high-frequency ones.

Get Help: Interactive Example – AC 1
P'Cast 21.6 – Tuning an FM Radio

Chapter 22
Electromagnetic Waves

Conceptual Questions

22.3 The electric field is perpendicular to the magnetic field and the velocity of the wave. That means that the electric fields will vibrate, say, along the x axis, the magnetic field along the x axis, and the EM radiation will move along the z axis. These three vectors are mutually perpendicular.

Get Help: P'Cast 22.1 – A Radio Wave

22.7 All electromagnetic waves travel at the same speed in a vacuum: $c = 3.00 \times 10^8$ m/s. But the energy of an individual photon is proportional to the frequency of that wave: $E = hf$. Therefore, to transmit energy at a given rate, a relatively high-frequency wave will transmit fewer photons per second than a relatively low-frequency wave.

Get Help: P'Cast 22.3 – A Magnetic Field Due to a Changing Electric Field

Multiple-Choice Questions

22.11 C (have same speed). All EM radiation travels at the speed of light.

Get Help: P'Cast 22.1 – A Radio Wave

22.13 D (sound). EM waves can travel in a vacuum.

Get Help: P'Cast 22.1 – A Radio Wave

22.15 E (to both time-independent and time-dependent electric and magnetic fields). Maxwell's equations are the fundamental relationships underlying all electric, magnetic, and electromagnetic phenomena, regardless of their time dependence.

Get Help: P'Cast 22.3 – A Magnetic Field Due to a Changing Electric Field

Estimation/Numerical Analysis

22.19 Green light has a wavelength around 550 nm. The energy of a green photon is then

$$E = hf$$

$$c = f\lambda$$

$$E = \frac{hc}{\lambda} = \frac{(6.63 \times 10^{-34} \text{ J} \cdot \text{s})\left(3.00 \times 10^8 \frac{\text{m}}{\text{s}}\right)}{550 \times 10^{-9} \text{ m}} = 3.6 \times 10^{-19} \text{ J}$$

22.23 The people 300 km away receive the news first:

$$v = \frac{\Delta x}{\Delta t}$$

$$t_{radio} = \frac{\Delta x}{c} = \frac{300 \times 10^3 \text{ m}}{\left(3.00 \times 10^8 \frac{\text{m}}{\text{s}}\right)} = 1 \times 10^{-3} \text{ s}$$

$$t_{sound} = \frac{\Delta x}{v_{sound}} = \frac{3 \text{ m}}{\left(340 \frac{\text{m}}{\text{s}}\right)} = 9 \times 10^{-3} \text{ s}$$

Get Help: P'Cast 22.1 – A Radio Wave

Problems

22.27

SET UP

We are given a list of the wavelengths of some photons and asked to calculate their frequencies. The frequency of a photon is related to the wavelength by the speed of light, $c = f\lambda$.

SOLVE

$$c = f\lambda$$
$$f = \frac{c}{\lambda}$$

A) $f = \dfrac{c}{\lambda} = \dfrac{\left(3.00 \times 10^8 \frac{\text{m}}{\text{s}}\right)}{700 \times 10^{-9} \text{ m}} = \boxed{4.29 \times 10^{14} \text{ Hz}}$

B) $f = \dfrac{c}{\lambda} = \dfrac{\left(3.00 \times 10^8 \frac{\text{m}}{\text{s}}\right)}{600 \times 10^{-9} \text{ m}} = \boxed{5.00 \times 10^{14} \text{ Hz}}$

C) $f = \dfrac{c}{\lambda} = \dfrac{\left(3.00 \times 10^8 \frac{\text{m}}{\text{s}}\right)}{500 \times 10^{-9} \text{ m}} = \boxed{6.00 \times 10^{14} \text{ Hz}}$

D) $f = \dfrac{c}{\lambda} = \dfrac{\left(3.00 \times 10^8 \frac{\text{m}}{\text{s}}\right)}{400 \times 10^{-9} \text{ m}} = \boxed{7.50 \times 10^{14} \text{ Hz}}$

E) $f = \dfrac{c}{\lambda} = \dfrac{\left(3.00 \times 10^8 \frac{\text{m}}{\text{s}}\right)}{100 \times 10^{-9} \text{ m}} = \boxed{3.00 \times 10^{15} \text{ Hz}}$

58   Chapter 22 Electromagnetic Waves

F) $f = \dfrac{c}{\lambda} = \dfrac{\left(3.00 \times 10^8 \dfrac{m}{s}\right)}{0.0333 \times 10^{-9} \text{ m}} = \boxed{9.01 \times 10^{18} \text{ Hz}}$

G) $f = \dfrac{c}{\lambda} = \dfrac{\left(3.00 \times 10^8 \dfrac{m}{s}\right)}{500 \times 10^{-6} \text{ m}} = \boxed{6.00 \times 10^{11} \text{ Hz}}$

H) $f = \dfrac{c}{\lambda} = \dfrac{\left(3.00 \times 10^8 \dfrac{m}{s}\right)}{63.3 \times 10^{-12} \text{ m}} = \boxed{4.74 \times 10^{18} \text{ Hz}}$

REFLECT
The frequency and wavelength are inversely proportional to one another, so a smaller wavelength will have a larger frequency, and vice versa.

Get Help: P'Cast 22.1 – A Radio Wave

22.31

SET UP
The speed of light in a vacuum is $c = 3.00 \times 10^8$ m/s. The distance light travels in 10×10^{-9} s can be found by multiplying the speed by the time interval.

SOLVE

$$c = \dfrac{\Delta x}{\Delta t}$$

$$\Delta x = c(\Delta t) = \left(3.00 \times 10^8 \dfrac{m}{s}\right)(10 \times 10^{-9} \text{ s}) = \boxed{3 \text{ m}}$$

REFLECT
One light-year is the distance light travels during one year: 1 ly = 9.4607×10^{12} km.

Get Help: P'Cast 22.1 – A Radio Wave

22.35

SET UP
We can rewrite the speed of light in terms of the angular frequency ω and the wave number k, $c = \dfrac{\omega}{k}$, in order to solve for the wave number of a photon with $\omega = 6.28 \times 10^{15} \dfrac{\text{rad}}{\text{s}}$ or the angular frequency of a photon with $k = 4\pi \times 10^6 \dfrac{\text{rad}}{\text{m}}$. The frequency is equal to $f = \dfrac{\omega}{2\pi}$, and the wavenumber is related to the wavelength though its definition, $k = \dfrac{2\pi}{\lambda}$.

SOLVE
Speed of light in terms of ω and k:

$$c = f\lambda = \left(\dfrac{\omega}{2\pi}\right)\left(\dfrac{2\pi}{k}\right) = \left(\dfrac{\omega}{k}\right)$$

Part a)

$$c = \frac{\omega}{k}$$

$$k = \frac{\omega}{c} = \frac{\left(6.28 \times 10^{15} \frac{\text{rad}}{\text{s}}\right)}{\left(3.00 \times 10^{8} \frac{\text{m}}{\text{s}}\right)} = \boxed{2.09 \times 10^{7} \frac{\text{rad}}{\text{m}}}$$

Part b)

Angular frequency:

$$c = \frac{\omega}{k}$$

$$\omega = ck = \left(3.00 \times 10^{8} \frac{\text{m}}{\text{s}}\right)\left(4\pi \times 10^{6} \frac{\text{rad}}{\text{m}}\right) = \boxed{3.77 \times 10^{15} \frac{\text{rad}}{\text{s}}}$$

Frequency:

$$f = \frac{\omega}{2\pi} = \frac{\left(3.77 \times 10^{15} \frac{\text{rad}}{\text{s}}\right)}{2\pi} = \boxed{6.00 \times 10^{14} \text{ Hz}}$$

Wavelength:

$$k = \frac{2\pi}{\lambda}$$

$$\lambda = \frac{2\pi}{k} = \frac{2\pi}{\left(4\pi \times 10^{6} \frac{\text{rad}}{\text{m}}\right)} = \boxed{5.00 \times 10^{-7} \text{ m}}$$

REFLECT

The photon in part (a) is 300-nm ultraviolet light. The photon in part (b) is blue-green visible light.

Get Help: P'Cast 22.1 – A Radio Wave

22.39

SET UP

Charge is flowing onto the positive plate and off of the negative plate at a rate of $\frac{\Delta q}{\Delta t} = 2.8$ A. If we treat the parallel plates as being approximately infinite, the electric field in between the plates has a magnitude of $E = \frac{q}{A\varepsilon_0}$, where A is the cross-sectional area. The displacement current through the capacitor between the plates is equal to $\varepsilon_0 \frac{\Delta \Phi_E}{\Delta t}$, where $\Phi_E = EA\cos\theta$.

SOLVE

$$\varepsilon_0 \frac{\Delta \Phi_E}{\Delta t} = \varepsilon_0 \frac{\Delta(EA\cos\theta)}{\Delta t} = \varepsilon_0 \frac{\Delta(EA)}{\Delta t} = \varepsilon_0 \frac{A\Delta\left(\frac{q}{A\varepsilon_0}\right)}{\Delta t} = \frac{\Delta q}{\Delta t} = \boxed{2.8 \text{ A}}$$

60 Chapter 22 Electromagnetic Waves

REFLECT

It makes sense that the displacement current "through" the capacitor should equal the rate of charge flowing onto the positive plate and off of the negative plate.

Get Help: P'Cast 22.3 – A Magnetic Field Due to a Changing Electric Field

General Problems

22.49

SET UP

A laser is made up of a cylindrical beam of diameter $d = 0.750 \times 10^{-2}$ m. The energy is pulsed, lasting $\Delta t = 1.50 \times 10^{-9}$ s. Each pulse contains an energy of $E = 2.00$ J. The length (in meters) of the pulse can be found by multiplying the pulse duration by the speed of light c. Using this length, we can calculate the volume of the cylindrical pulse and then the energy per unit volume for each pulse.

SOLVE

Part a)

$$v = \frac{\Delta x}{\Delta t} = c$$

$$\Delta x = c\Delta t = \left(3.00 \times 10^8 \frac{\text{m}}{\text{s}}\right)(1.50 \times 10^{-9} \text{ s}) = \boxed{0.450 \text{ m}}$$

Part b)

$$\frac{E}{V} = \frac{E}{\pi r^2 (\Delta x)} = \frac{E}{\pi \left(\frac{d}{2}\right)^2 (\Delta x)} = \frac{4E}{\pi d^2 (\Delta x)} = \frac{4(2.00 \text{ J})}{\pi (0.750 \times 10^{-2} \text{ m})^2 (0.450 \text{ m})} = \boxed{1.01 \times 10^5 \frac{\text{J}}{\text{m}^3}}$$

REFLECT

As a rule of thumb, light travels a distance of 1 m in 3.3 ns.

Get Help: P'Cast 22.4 – Energy Density and Intensity in a Radio Wave

Chapter 23
Wave Properties of Light

Conceptual Questions

23.3 If Earth had no atmosphere, Earth would completely block all sunlight from reaching the Moon during a lunar eclipse, and the Moon would be invisible. However, because of Earth's atmosphere, light is scattered into the shadow cast on the Moon by Earth. Short wavelengths of light are more effectively scattered away by the atmosphere, but longer wavelengths, such as red and orange, will not be scattered as much and will reach the Moon's surface. Thus, the Moon will appear red during a lunar eclipse.

Get Help: Interactive Example – Refraction
P'Cast 23.1 – Seeing under Water

23.9 When light enters a material with a negative index of refraction, the light refracts "back" away from the normal to the surface as seen in the picture below:

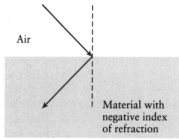

Figure 23-1 Problem 9

Get Help: Interactive Example – Refraction
P'Cast 23.1 – Seeing under Water

23.15 Sunlight includes all of the colors of the rainbow, and one of the delightful aspects of a diamond is its strong dispersion. Each color of visible light, corresponding to a different range of wavelengths, will bend at different angles through a diamond so that sunlight is separated into its different colors as it passes through the diamond's surface.

Get Help: Interactive Example – Refraction
P'Cast 23.1 – Seeing under Water

Multiple-Choice Questions

23.21 B $\left(\frac{1}{2}I_0\right)$. The intensity of unpolarized light drops by a factor of two when it passes through a linear polarizer. The second polarizer does not affect the intensity of the light because the two polarizers are aligned (*i.e.*, $\theta = 0$).

Get Help: Interactive Example – Polarization I
P'Cast 23.5 – Brewster's Angle for Air to Water

23.25 A ($w_a > w_b$). The width of the slit is inversely proportional to the fringe width.

(a) (b)

Figure 23-2 Problem 25

Get Help: Interactive Example – Missing Order
P'Cast 23.8 – Diffraction Through a Slit

Estimation/Numerical Questions

23.29 We can use $c = \dfrac{\Delta x}{\Delta t}$ to calculate the distance light travels.

Part a)

$$\Delta x = c(\Delta t) = \left(3.00 \times 10^8 \dfrac{\text{m}}{\text{s}}\right)(1 \text{ s}) = 3 \times 10^8 \text{ m}$$

Part b)

$$\Delta x = c(\Delta t) = \left(3.00 \times 10^8 \dfrac{\text{m}}{\text{s}}\right)(60 \text{ s}) = 2 \times 10^{10} \text{ m}$$

Part c)

$$\Delta x = c(\Delta t) = \left(3.00 \times 10^8 \dfrac{\text{m}}{\text{s}}\right)\left(1 \text{ y} \times \dfrac{365.25 \text{ days}}{1 \text{ y}} \times \dfrac{24 \text{ h}}{1 \text{ day}} \times \dfrac{3600 \text{ s}}{1 \text{ h}}\right) \approx 9 \times 10^{15} \text{ m}$$

Get Help: Interactive Example – Refraction
P'Cast 23.1 – Seeing under Water

Problems

23.35

SET UP

The speed of light in methylene iodide, abbreviated MeI, is $v_{\text{MeI}} = 1.72 \times 10^8 \dfrac{\text{m}}{\text{s}}$. The index of refraction of water is $n_{\text{water}} = 1.33$, which means the speed of light in water is $v_{\text{water}} = \dfrac{c}{n_{\text{water}}}$. We can use these data along with the definition of speed, $v = \dfrac{\Delta x}{\Delta t}$, to calculate the distance of methylene iodide that light must travel through such that it takes the same amount of time as light traveling through 1000×10^3 m of water.

SOLVE

$$v = \dfrac{\Delta x}{\Delta t}$$

$$\Delta t = \frac{\Delta x}{v}$$

$$\frac{\Delta x_{\text{MeI}}}{v_{\text{MeI}}} = \frac{\Delta x_{\text{water}}}{v_{\text{water}}}$$

$$\Delta x_{\text{MeI}} = \frac{\Delta x_{\text{water}} v_{\text{MeI}}}{v_{\text{water}}} = \frac{\Delta x_{\text{water}} v_{\text{MeI}}}{\left(\frac{c}{n_{\text{water}}}\right)} = \frac{\Delta x_{\text{water}} v_{\text{MeI}} n_{\text{water}}}{c}$$

$$= \frac{(1000 \times 10^3 \text{ m})\left(1.72 \times 10^8 \frac{\text{m}}{\text{s}}\right)(1.33)}{\left(3.00 \times 10^8 \frac{\text{m}}{\text{s}}\right)} = \boxed{7.63 \times 10^5 \text{ m} = 763 \text{ km}}$$

REFLECT

The index of refraction of methylene iodide is $n_{\text{MeI}} = \frac{c}{v_{\text{MeI}}} = \frac{\left(3.00 \times 10^8 \frac{\text{m}}{\text{s}}\right)}{\left(1.72 \times 10^8 \frac{\text{m}}{\text{s}}\right)} = 1.74$, which means light will travel more slowly in methylene iodide compared to water. Therefore, we expect that the distance the light travels in methylene iodide should be smaller than the distance in water.

Get Help: Interactive Example – Refraction
P'Cast 23.1 – Seeing under Water

23.39

SET UP

Total internal reflection occurs when the incident angle of light traveling from a medium with index of refraction n_1 into a medium with index of refraction n_2 is greater than the critical angle $\theta_c = \sin^{-1}\left(\frac{n_2}{n_1}\right)$. It can only occur when $n_1 > n_2$. We can use the provided indices of refraction to calculate the critical angle in each case.

SOLVE

A) $\theta_c = \sin^{-1}\left(\frac{n_2}{n_1}\right) = \sin^{-1}\left(\frac{n_{\text{air}}}{n_{\text{plastic}}}\right) = \sin^{-1}\left(\frac{1.00}{1.50}\right) = \boxed{41.8°}$

B) $\theta_c = \sin^{-1}\left(\frac{n_2}{n_1}\right) = \sin^{-1}\left(\frac{n_{\text{air}}}{n_{\text{water}}}\right) = \sin^{-1}\left(\frac{1.00}{1.33}\right) = \boxed{48.8°}$

C) $\theta_c = \sin^{-1}\left(\frac{n_2}{n_1}\right) = \sin^{-1}\left(\frac{n_{\text{water}}}{n_{\text{glass}}}\right) = \sin^{-1}\left(\frac{1.33}{1.56}\right) = \boxed{58.5°}$

D) $\theta_c = \sin^{-1}\left(\frac{n_2}{n_1}\right) = \sin^{-1}\left(\frac{n_{\text{glass}}}{n_{\text{air}}}\right) = \sin^{-1}\left(\frac{1.55}{1.00}\right) = \boxed{\text{no critical angle}}$

64 Chapter 23 Wave Properties of Light

REFLECT

Because $-1 \leq \sin(\theta) \leq 1$, $\sin^{-1}\left(\dfrac{1.55}{1.00}\right) = \sin^{-1}(1.55)$ is undefined.

Get Help: Picture It – Refraction and Reflection
P'Cast 23.2 – Critical Angles
P'Cast 23.3 – Laser Lithotripsy

23.43

SET UP

A scuba diver wants to look up from inside the water ($n_{water} = 1.33$) to see her friend standing on a very distant shore ($n_{air} = 1.00$). Since her friend is located so far away, the angle of the light ray is essentially 90°. We can use Snell's law to calculate the angle the scuba diver must look.

SOLVE

$$n_1 \sin\theta_1 = n_2 \sin\theta_2$$

$$\sin\theta_1 = \frac{n_2}{n_1}\sin\theta_2$$

$$\theta_1 = \sin^{-1}\left(\frac{n_{air}}{n_{water}}\sin\theta_2\right) = \sin^{-1}\left(\frac{1.00}{1.33}\sin 90°\right) = 48.8°$$

REFLECT

This is equivalent to finding the critical angle for the system.

Get Help: Picture It – Refraction and Reflection
P'Cast 23.2 – Critical Angles
P'Cast 23.3 – Laser Lithotripsy

23.49

SET UP

Blue light and yellow light are incident on a glass slab ($t = $ 12 cm) at an angle of 45° relative to the normal. The index of refraction for the blue light in the glass is $n_{glass,blue} = 1.545$, and the index of refraction for the yellow light in the glass is $n_{glass,yellow} = 1.523$. We are interested in the distance separating the two rays when they emerge from the other side of the slab. In order to calculate this, we first need to find the vertical displacement of each ray (either y_{blue} or y_{yellow} for blue or yellow light, respectively) once it exits the slab. Applying Snell's law at the first interface will give the angle of the refracted light (either θ_{blue} or θ_{yellow} for blue or yellow light, respectively) in the glass slab. The ray will travel at this angle until it reaches the other side. We can draw a triangle and relate the thickness of the slab t and the tangent of θ_{blue} (or θ_{yellow}) to the vertical displacement y_{blue} (or y_{yellow}). The distance between the rays is equal to the difference between the two vertical displacements.

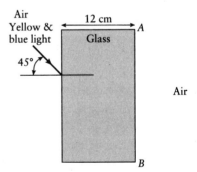

Figure 23-3 Problem 49

SOLVE

Blue light, angle of refraction:

$$n_{air}\sin(\theta_1) = n_{glass,blue}\sin(\theta_{blue})$$

$$\theta_{blue} = \sin^{-1}\left(\frac{n_{air}}{n_{glass,blue}}\sin(\theta_1)\right) = \sin^{-1}\left(\left(\frac{1.000}{1.545}\right)\sin(45°)\right) = 27.237°$$

Blue light, vertical displacement:

Figure 23-4 Problem 49

$$\tan(\theta_{blue}) = \frac{y_{blue}}{t}$$

$$y_{blue} = t\tan(\theta_{blue}) = (12\text{ cm})\tan(27.237°) = 6.1770\text{ cm}$$

Yellow light, angle of refraction:

$$n_{air}\sin(\theta_1) = n_{glass,yellow}\sin(\theta_{yellow})$$

$$\theta_{yellow} = \sin^{-1}\left(\frac{n_{air}}{n_{glass,yellow}}\sin(\theta_1)\right) = \sin^{-1}\left(\left(\frac{1.000}{1.523}\right)\sin(45°)\right) = 27.664°$$

Yellow light, vertical displacement:

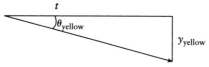

Figure 23-5 Problem 49

$$\tan(\theta_{yellow}) = \frac{y_{yellow}}{t}$$

$$y_{yellow} = t\tan(\theta_{yellow}) = (12\text{ cm})\tan(27.664°) = 6.2905\text{ cm}$$

Vertical difference between the blue and yellow rays:

$$\Delta y = y_{yellow} - y_{blue} = (6.2905\text{ cm}) - (6.1770\text{ cm}) = \boxed{0.1135\text{ cm}}$$

REFLECT

The thicker the piece of glass, the farther apart the blue and yellow rays will emerge from one another, which makes sense. The shallower the angle of the incoming light is, the closer together the rays will emerge, which also makes sense.

23.51

SET UP

Vertically polarized light with an initial intensity of I_0 is sent through a polarizing filter that makes a relative angle of θ to the filter. The intensity of the light after the polarizer is reduced by 25% such that $I = 0.75 I_0$. The intensity of polarized light after it passes through a linear polarizer that makes a relative angle of θ is equal to $I = I_0 \cos^2(\theta)$.

SOLVE

$$I = I_0 \cos^2 \theta$$

$$\theta = \cos^{-1}\left(\pm\sqrt{\frac{I}{I_0}}\right) = \cos^{-1}\left(\pm\sqrt{\frac{0.75 I_0}{I_0}}\right) = \cos^{-1}(\pm\sqrt{0.75}) = \boxed{30°, 150°}$$

REFLECT

It makes sense that there should be two angles due to the symmetry of the system.

Get Help: Interactive Example – Polarization I
P'Cast 23.5 – Brewster's Angle for Air to Water

23.57

SET UP

A person observes that the light rays from the Sun that bounce off the air–water surface are linearly polarized along the horizontal. When incoming light strikes the air–water boundary at Brewster's angle, $\theta_B = \tan^{-1}\left(\frac{n_{\text{water}}}{n_{\text{air}}}\right)$, the reflected light is completely polarized parallel to the surface. Brewster's angle is measured relative to the vertical, so the angle relative to the horizontal is equal to $\theta = 90.0° - \theta_B$.

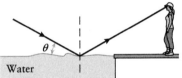

Figure 23-6 Problem 57

SOLVE

Brewster's angle:

$$\theta_B = \tan^{-1}\left(\frac{n_2}{n_1}\right) = \tan^{-1}\left(\frac{n_{\text{water}}}{n_{\text{air}}}\right) = \tan^{-1}\left(\frac{1.33}{1.00}\right) = 53.1°$$

Angle above the horizontal:

$$\theta = 90.0° - \theta_B = 90.0° - 53.1° = \boxed{36.9°}$$

REFLECT

If the Sun is close to, but not exactly at, this angle, the reflected light will be strongly, not completely, polarized.

Get Help: Interactive Example – Polarization I
P'Cast 23.5 – Brewster's Angle for Air to Water

23.61

SET UP

White light illuminates a thin film ($n_{film} = 1.35$) normal to the surface, and we observe that both blue light ($\lambda_{blue} = 500$ nm) and red light ($\lambda_{red} = 700$ nm) are strongly reflected. The film has a thickness t and is floating on top of water ($n_{water} = 1.33$). Since the film has a higher index of refraction than water, constructive interference is described by the following equation: $2D = (2m - 1)\dfrac{n_1 \lambda_1}{2n_2}$. This same path difference strongly reflects two wavelengths of light, so we will have two expressions with different integers, m_{blue} for the blue light and m_{red} for the red light. Setting the thicknesses equal to one another, we can solve the smallest integers that satisfy the resulting relationship. The minimum thickness of the film can be found by plugging the integer back into the constructive interference relationship.

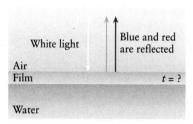

Figure 23-7 Problem 61

SOLVE

$$2D = (2m - 1)\frac{n_1 \lambda_1}{2n_2}$$

Blue light:

$$2D = (2m_{blue} - 1)\frac{n_{air} \lambda_{blue}}{2n_{film}}$$

Red light:

$$2D = (2m_{red} - 1)\frac{n_{air} \lambda_{red}}{2n_{film}}$$

Solving for m's:

$$(2m_{blue} - 1)\frac{n_{air} \lambda_{blue}}{2n_{film}} = (2m_{red} - 1)\frac{n_{air} \lambda_{red}}{2n_{film}}$$

$$(2m_{blue} - 1)\lambda_{blue} = (2m_{red} - 1)\lambda_{red}$$

$$(2m_{blue} - 1)(500 \text{ nm}) = (2m_{red} - 1)(700 \text{ nm})$$

$$1400 m_{red} - 1000 m_{blue} = 200$$

$$7 m_{red} - 5 m_{blue} = 1$$

This is true when $m_{red} = 3$ and $m_{blue} = 4$.

Thickness:

$$2D = (2m_{blue} - 1)\frac{n_{air} \lambda_{blue}}{2n_{film}}$$

$$D = (2(4) - 1)\frac{(1.00)(500 \text{ nm})}{4(1.35)}$$

$$D = 648 \text{ nm} \approx \boxed{650 \text{ nm}}$$

REFLECT

Since we are looking for the minimum thickness of the film, we chose the smallest values of m_{blue} and m_{red} that satisfied the equation $7m_{red} - 5m_{blue} = 1$. The next set of integers that satisfies that relationship is $m_{red} = 8$ and $m_{blue} = 11$. This corresponds to a thickness of approximately 1900 nm.

Get Help: Interactive Example – Film on Water
P'Cast 23.6 – A Soapy Film
P'Cast 23.7 – Reducing the Reflection

23.65

SET UP

A pool of water ($n_{water} = 1.33$) covered with a thin film of oil ($D = 450$ nm, $n_{oil} = 1.45$) is illuminated with white light and viewed from straight above. We want to know which visible wavelengths are *not* present in the reflected light, which means these wavelengths undergo total destructive interference. Since the film has a higher index of refraction than the water, destructive interference is described by the following equation: $2D = m\dfrac{n_1 \lambda_1}{n_2}$. We can solve for an expression for the wavelength in terms of the integer m. This will allow us to plug in consecutive values for m (starting from $m = 1$) and determine which wavelengths lie in the visible region (around 380 nm–750 nm)

SOLVE

Solving for wavelength:

$$2D = m\dfrac{n_1 \lambda_1}{n_2}$$

$$\lambda = \dfrac{2Dn_2}{mn_1}$$

For $m = 1$:

$$\lambda = \dfrac{2Dn_2}{mn_1} = \dfrac{2(450 \text{ nm})(1.45)}{(1)(1.00)} = 1305 \text{ nm}$$

1305 nm is in the infrared region, so we are not interested in this wavelength.
For $m = 2$:

$$\lambda = \dfrac{2Dn_2}{mn_1} = \dfrac{2(450 \text{ nm})(1.45)}{(2)(1.00)} = \boxed{652 \text{ nm}}$$

652 nm corresponds with red/orange visible light.
For $m = 3$:

$$\lambda = \dfrac{2Dn_2}{mn_1} = \dfrac{2(450 \text{ nm})(1.45)}{(3)(1.00)} = \boxed{435 \text{ nm}}$$

435 nm corresponds with violet/indigo visible light.
For $m = 4$:

$$\lambda = \dfrac{2Dn_2}{mn_1} = \dfrac{2(450 \text{ nm})(1.45)}{(4)(1.00)} = 326 \text{ nm}$$

326 nm is in the ultraviolet region, so we are not interested in this wavelength.

REFLECT
The visible wavelength that is strongly reflected is 522 nm (green).

> **Get Help:** Interactive Example – Film on Water
> P'Cast 23.6 – A Soapy Film
> P'Cast 23.7 – Reducing the Reflection

23.71

SET UP
The beam from a He-Ne laser illuminates a single slit of width $w = 1850$ nm. In the resulting diffraction pattern, the first dark fringe appears at an angle of 20.0° from the central maximum. We can use the expression for the angular position of the first dark fringe, $\sin(\theta) = \frac{\lambda}{w}$, to calculate the wavelength of the laser light.

SOLVE

$$\sin(\theta) = \frac{\lambda}{w}$$

$$\lambda = w\sin(\theta) = (1850 \text{ nm})\sin(20.0°) = \boxed{633 \text{ nm}}$$

REFLECT
A typical He-Ne laser appears red, and a wavelength of 633 nm is well within the range of red visible light.

> **Get Help:** Interactive Example – Missing Order
> P'Cast 23.8 – Diffraction Through a Slit

23.77

SET UP
Light of wavelength λ is sent through a circular aperture of diameter D. The diffraction pattern is projected on a screen located a distance $L = 0.85$ m from the aperture. The first dark ring is 15,000λ from the center of the central maximum. The angular position of the first dark ring is given by $\sin(\theta) = 1.22\frac{\lambda}{D}$. Using the small angle approximation, we can relate this to $\tan(\theta)$ by geometry. Putting all of this together will allow us to calculate the diameter of the aperture D.

SOLVE
Geometry:

Figure 23-8 Problem 77

Diameter of aperture:

$$\sin(\theta) = 1.22\frac{\lambda}{D}$$

Using the small angle approximation, $\sin(\theta) \approx \tan(\theta) = \dfrac{15{,}000\lambda}{L}$:

$$\dfrac{15{,}000\lambda}{L} = 1.22\dfrac{\lambda}{D}$$

$$D = \dfrac{1.22 L}{15{,}000} = \dfrac{1.22(0.85\ \text{m})}{15{,}000} = \boxed{7 \times 10^{-5}\ \text{m}}$$

REFLECT

The wavelength of visible light is on the order of hundreds of nanometers. If we assume 5×10^{-7} m as a typical value, the first dark ring is 7.5×10^{-3} m from the center of the central maximum. This is approximately 2 orders of magnitude smaller than the distance between the aperture and the screen, so our use of the small angle approximation is justified.

Get Help: P'Cast 23.9 – The Hubble Space Telescope

23.83

SET UP

A flat glass surface ($n_{\text{glass}} = 1.54$) has a uniform layer of water ($n_{\text{water}} = 1.33$) on top of the glass. We want to know the minimum angle of incidence that light coming from the glass must strike the glass–water interface such that the light is totally internally reflected by the water–air interface. The easiest way of tackling this problem is by working backwards. First, we can calculate the critical angle for the water–air interfaces from $\theta_c = \sin^{-1}\left(\dfrac{n_2}{n_1}\right)$. This angle is also equal to the angle at which the light is refracted by the glass–water interface, which means we can apply Snell's law and calculate the incident angle of the light in the glass.

SOLVE

Critical angle for water–air interface:

$$\theta_c = \sin^{-1}\left(\dfrac{n_2}{n_1}\right) = \sin^{-1}\left(\dfrac{n_{\text{air}}}{n_{\text{water}}}\right) = \sin^{-1}\left(\dfrac{1.00}{1.33}\right) = 48.75°$$

Angle of incidence for the glass–water interface:

$$n_1 \sin\theta_1 = n_2 \sin\theta_2$$

$$n_{\text{glass}} \sin\theta_1 = n_{\text{water}} \sin\theta_c$$

$$\theta_1 = \sin^{-1}\left(\dfrac{n_{\text{water}}}{n_{\text{glass}}} \sin(\theta_c)\right) = \sin^{-1}\left(\dfrac{1.33}{1.54} \sin(48.75°)\right) = \boxed{40.5°}$$

REFLECT

We could have saved ourselves an intermediate step by realizing the water layer effectively does not enter into the calculation. Applying Snell's law at the first interface, we find $n_{\text{glass}} \sin(\theta_1) = n_{\text{water}} \sin(\theta_2)$. Applying it again at the second interface we get $n_{\text{water}} \sin(\theta_2) = n_{\text{air}} \sin(90°)$. Combining these two expressions, we can eliminate the term related to the water and see that $n_{\text{glass}} \sin(\theta_1) = n_{\text{air}}$, or

$$\theta_1 = \sin^{-1}\left(\dfrac{n_{\text{air}}}{n_{\text{glass}}}\right) = \sin^{-1}\left(\dfrac{1.00}{1.54}\right) = 40.5°.$$

23.89

Get Help: Picture It – Refraction and Reflection
P'Cast 23.2 – Critical Angles
P'Cast 23.3 – Laser Lithotripsy

SET UP

Unpolarized light with an intensity of $I_0 = 100\frac{W}{m^2}$ is incident on three polarizers. Two of them are placed with their transmission axes perpendicular to each other. A third polarizer is placed in between the two crossed polarizers such that the transmission axis of the second polarizer is oriented 30° relative to that of the first. This means the transmission axis of the third polarizer is oriented 60° relative to that of the second. In general, the intensity of unpolarized light drops by a factor of two when it passes through a linear polarizer. The intensity of the now linearly polarized light after it passes through a polarizer depends on the previous intensity and the square of the cosine of the angle between the polarization axis of the light and the transmission axis of the polarizer. The orientation of the middle polarizer that maximizes the transmitted intensity can be found by analyzing the maximum value of the sinusoidal function describing intensity. Since the first and third polarizers are perpendicular to one another, we can call the angle between polarizers 1 and 2 θ and the angle between polarizers 2 and 3 $(90° - \theta)$.

SOLVE

$$I = I_0 \cos^2\theta$$

Part a)

$$I_1 = \frac{I_0}{2}$$

$$I_2 = I_1 \cos^2(30°) = \left(\frac{I_0}{2}\right)\left(\frac{\sqrt{3}}{2}\right)^2 = \frac{3I_0}{8}$$

$$I_3 = I_2 \cos^2(60°) = \left(\frac{3I_0}{8}\right)\left(\frac{1}{2}\right)^2 = \frac{3I_0}{32} = \frac{3}{32}\left(100\frac{W}{m^2}\right) = \boxed{9\frac{W}{m^2}}$$

Part b)

Finding I_3 in terms of the general angle θ:

$$I_2 = I_1 \cos^2(\theta) = \frac{I_0}{2}\cos^2(\theta)$$

$$I_3 = I_2 \cos^2(90° - \theta) = \left(\frac{I_0}{2}\cos^2(\theta)\right)(\sin^2(\theta)) = \frac{I_0}{2}(\sin(\theta)\cos(\theta))^2$$

$$= \frac{I_0}{2}\left(\frac{\sin(2\theta)}{2}\right)^2 = \frac{I_0}{8}\sin^2(2\theta)$$

72 Chapter 23 Wave Properties of Light

Finding the maximum:
The function will be a maximum where $\sin^2(2\theta) = 1$. Note the following trigonometric identity:

$$\sin^2(2\theta) = \frac{1}{2} - \frac{1}{2}\cos(4\theta)$$

Setting this equal to 1:

$$\frac{1}{2} - \frac{1}{2}\cos(4\theta) = 1$$

$$\frac{1}{2}\cos(4\theta) = -\frac{1}{2}$$

$$\cos(4\theta) = -1$$

$$4\theta = \cos^{-1}(-1) = 180°$$

$$\theta = 45°$$

REFLECT
After the fact, it makes sense that the middle sheet should be at an angle of 45° relative to each polarizer due to symmetry in order to maximize the transmitted intensity.

Get Help: Interactive Example – Polarization I
P'Cast 23.5 – Brewster's Angle for Air to Water

23.93

SET UP
A slip of paper of thickness T is placed between the edges of two thin plates of glass that have a length of $L = 0.125$ m. This causes the top plate to make an angle θ with respect to the bottom plate. When light ($\lambda = 600 \times 10^{-9}$ m) is shone normally on the glass plates, interference fringes due to destructive interference are observed. The spacing along the plate between neighboring fringes is $x = 0.200 \times 10^{-3}$ m. The thickness of the air film changes along the length of the plates. We can use the relationship for destructive interference in order to calculate the difference in the thickness of the film at successive fringes; this distance is related to x by $\sin(\theta)$. Since we expect the thickness of one sheet of paper to be very small, the angle between the glass plates will also be very small, which means we can invoke the small angle approximation: $\tan(\theta) \approx \sin(\theta) \approx \theta$. This angle is also related to the thickness of the paper and the length of the bottom plate by $\tan(\theta)$, which is approximately equal to θ. By setting the two expressions for θ equal to one another, we can solve for T.

SOLVE

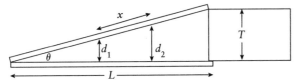

Figure 23-9 Problem 93

Destructive interference at successive fringes:

$$2D = m\lambda_{air}$$

First fringe:

$$2d_1 = m_1\lambda_{air}$$

Second fringe:

$$2d_2 = m_2\lambda_{air} = (m_1 + 1)\lambda_{air}$$

Horizontal distance between successive fringes:

$$2d_2 - 2d_1 = (m_1 + 1)\lambda_{air} - m_1\lambda_{air} = \lambda_{air}$$

$$d_2 - d_1 = \frac{\lambda_{air}}{2}$$

Angle in terms of the distance along the plate between successive fringes:

$$\sin(\theta) = \frac{d_2 - d_1}{x} = \frac{\left(\frac{\lambda_{air}}{2}\right)}{x} = \frac{\lambda_{air}}{2x}$$

$$\sin(\theta) \approx \theta \approx \frac{\lambda_{air}}{2x}$$

Angle in terms of the thickness of the paper:

$$\tan(\theta) = \frac{T}{L}$$

$$\tan(\theta) \approx \theta \approx \frac{T}{L}$$

Thickness of the paper:

$$\frac{T}{L} = \frac{\lambda_{air}}{2x}$$

$$T = \frac{\lambda_{air} L}{2x} = \frac{(600 \times 10^{-9} \text{ m})(0.125 \text{ m})}{2(0.200 \times 10^{-3} \text{ m})} = \boxed{1.88 \times 10^{-4} \text{ m} \approx 200 \ \mu m}$$

REFLECT

A thickness of about 0.2 mm seems reasonable for a single sheet of paper.

Get Help: Interactive Example – Missing Order
P'Cast 23.8 – Diffraction Through a Slit

23.97

SET UP

The distance between the crest and adjacent trough of a tsunami wave is 250 mi, which means the wavelength will be twice this distance. The period of the tsunami wave is $T = 1$ h. The speed of the wave is equal to the wavelength divided by the period. The time it takes the tsunami to travel a distance of $\Delta x = 600$ mi is equal to that distance divided by the wave speed. These waves travel between a 100-mile-wide opening. To determine whether we can use the diffraction relationship $w\sin(\theta) = m\lambda$, we need to compare the slit width w to the wavelength λ. The slit needs to be "narrow," which means w must be approximately the same size or smaller than the wavelength.

SOLVE

Part a)

$$v = \lambda f = \frac{\lambda}{T} = \frac{2(250 \text{ mi})}{1 \text{ h}} = \boxed{500 \text{ mph}}$$

Part b)

$$v = \frac{\Delta x}{\Delta t}$$

$$\Delta t = \frac{\Delta x}{v} = \frac{600 \text{ mi}}{500 \text{ mph}} = \boxed{1.2 \text{ h} \approx 1 \text{ h (rounded to one significant figure)}}$$

Part c) The formula $w\sin(\theta) = m\lambda$ applies only if the wavelength is smaller than the slit width w. In this case $w = 100$ mi and $\lambda = 500$ mi, so the formula $\boxed{\text{would } not \text{ apply}}$.

REFLECT

If we were to apply $w\sin(\theta) = m\lambda$ to find the first diffraction minimum, we would find that $\sin(\theta) = \frac{\lambda}{w} = \frac{500 \text{ mi}}{100 \text{ mi}} = 5$. The sine of an angle can only yield a result between -1 and 1, so there is no solution to this equation. Physically, this means the first diffraction minimum does not exist, and, thus, diffraction does not occur.

Get Help: Interactive Example – Missing Order
P'Cast 23.8 – Diffraction Through a Slit

23.103

SET UP

An optical telescope can achieve an angular resolution of 1/4 arcsec $\left(1 \text{ arcsec} = \left(\frac{1}{3600}\right)^\circ = 4.85 \times 10^{-6} \text{ rad}\right)$. Rayleigh's criterion for the angular resolution through a circular aperture is $\sin(\theta_R) = 1.22\frac{\lambda}{D}$, where we're interested in the wavelengths of visible light. We will use a wavelength of 550×10^{-9} m to represent visible light. We can use this, along with the Rayleigh criterion and the desired resolution of 1/4 arcsec, to calculate the minimum value of D.

SOLVE

Part a)

$$\sin(\theta_R) = 1.22\frac{\lambda}{D}$$

$$D = 1.22\frac{\lambda}{\sin(\theta_R)} = 1.22\frac{\lambda}{\sin\left(\frac{1}{4}\text{arcsec} \times \frac{4.85 \times 10^{-6} \text{ rad}}{1 \text{ arcsec}}\right)} = (1.01 \times 10^6)\lambda$$

$$= (1.01 \times 10^6)(550 \times 10^{-9} \text{ m}) = 0.55 \text{ m} = 55 \text{ cm}$$

The minimum diameter aperture should be around $\boxed{55 \text{ cm}}$ to achieve a resolution of 1/4 arcsec.

Part b) Building a telescope with a diameter much larger than this won't improve the resolution significantly as long as you have to look through Earth's atmosphere. Telescopes are built larger than the diffraction-limited diameter for greater light collecting power, which allows you to see dimmer objects.

REFLECT

The Rayleigh criterion is just one rule of thumb used to determine whether or not two objects can be resolved through a circular aperture.

Get Help: P'Cast 23.9 – The Hubble Space Telescope

Chapter 24
Geometrical Optics

Conceptual Questions

24.1 Actually, a plane mirror does neither, but instead inverts objects back to front. If the mirror inverted right and left, then the object's right hand that points east would appear on the images as a right hand pointing toward the west. Because the image is inverted back to front, the object facing north is transformed into an image that faces south. Also, the object's right hand is transformed into a left hand in the image.

Get Help: Picture It – Spherical Mirror

24.7 Additional information is needed. In accordance with the lensmaker's equation, if the radius of curvature of the front surface is larger, it is a diverging lens; if the radius of curvature of the front surface is smaller, it is a converging lens.

Get Help: Picture It – Converging Lens

Multiple-Choice Questions

24.15 B (The height of the image stays the same and the image distance increases). For a plane mirror, the image distance equals the object distance, so the image distance will increase as the object distance increases. Also, the image height is equal to the object height; since the height of the object doesn't change, neither will the height of the image.

Get Help: Picture It – Spherical Mirror

24.19 B (real and inverted).

$$\frac{1}{d_O} + \frac{1}{d_I} = \frac{1}{f}$$

$$\frac{1}{d_I} = \frac{1}{f} - \frac{1}{d_O} = \frac{1}{\left(\frac{r}{2}\right)} - \frac{1}{r} = \frac{2}{r} - \frac{1}{r} = \frac{1}{r}$$

$$d_I = r$$

$$m = -\frac{d_I}{d_O} = -\frac{r}{r} = -1$$

A positive image distance means the image is real. A negative magnification indicates the image is inverted.

24.23 B (The objective lens is a short focal length, converging lens and the eyepiece functions as a simple magnifier). The real image from the objective lens of a compound microscope converges just inside the focal point of the eyepiece, which means the angular size of the overall image will be much larger than the angular size of the object.

> **Get Help:** Interactive Example – Focal Lens
> P'Cast 24.3 – Calculating Focal Length for a Convex Lens
> P'Cast 24.4 – Calculating Focal Length for a Concave Lens
> P'Cast 24.5 – An Image Made by a Convex Lens

Estimation/Numerical Analysis

24.27 A shiny spoon (f about 5 cm), a shiny salad bowl (f about 15 cm), and a reflector in a flashlight (f about 2 cm).

> **Get Help:** Interactive Example – Spherical Mirror
> P'Cast 24.1 – Images Made by a Concave Mirror

Problems

24.31

SET UP
Two flat mirrors are perpendicular to each other. An incoming beam of light makes an angle of 30° with respect to the first mirror. The beam will reflect off the mirror at an angle of 30° according to the law of reflection. The beam then makes an angle of 60° with respect to the second mirror, or 30° with respect to the normal of the second mirror. Again, applying the law of reflection, we see that the outgoing beam will make an angle of 30° with respect to the normal of the second mirror.

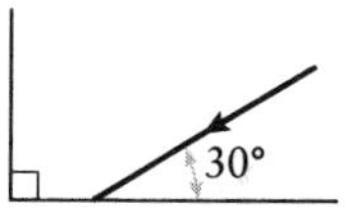

Figure 24-1 Problem 31

SOLVE

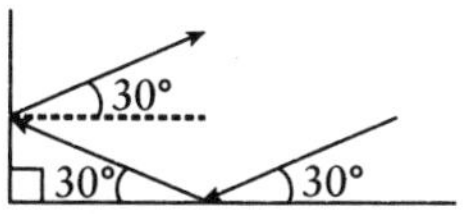

Figure 24-2 Problem 31

The outgoing beam will make an angle of $\boxed{30°}$ with respect to the normal of the second mirror.

REFLECT
Recognizing common geometric patterns will go a long way in solving optics problems. For example, the path of the ray leaving the first mirror and hitting the second mirror makes a 30-60-90 triangle with the two mirrors.

> **Get Help:** Picture It – Spherical Mirror

78 Chapter 24 Geometrical Optics

24.35

SET UP
A plane mirror is 10 m away from and parallel to a second plane mirror. An object is placed 3 m to the right of the left mirror, which means it is 7 m to the left of the right mirror. The original object creates an image in both the left and the right mirror. These images then act as objects for the other mirror and are thus reflected again. This process repeats and there will be an infinite number of images. In every case, the image distance is equal to the object distance for the mirror.

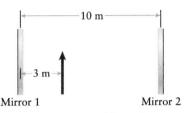

Figure 24-3 Problem 35

SOLVE
Illustration of the resulting images:

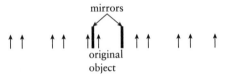

Figure 24-4 Problem 35

The first images formed in the left mirror are $\boxed{\text{3 m, 17 m, 23 m, 37 m, and 43 m to the left of the left mirror}}$.

The first images formed in the right mirror are $\boxed{\text{7 m, 13 m, 27 m, 33 m, and 47 m to the right of the right mirror}}$.

REFLECT
You can see this effect for yourself at the hair salon or in a mirror maze or even if you have two mirrors in your bathroom.

Get Help: Picture It – Spherical Mirror

24.41

SET UP
We can use ray tracing in order to determine if there are any situations where a real image is formed by a spherical, concave mirror.

SOLVE
An arrow is placed in front of a spherical, concave mirror. We will trace two light rays from the tip of the arrow to the mirror and as they reflect from it.

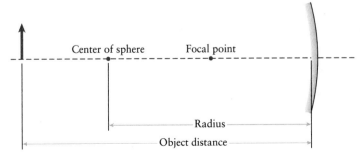

Figure 24-5 Problem 41

The image of the point of the arrow forms where the two rays cross. The image is inverted relative to the object.

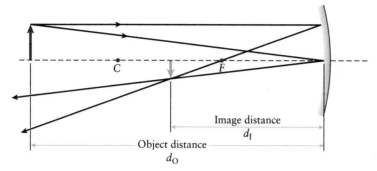

Figure 24-6 Problem 41

Two light rays are traced from the tip of the arrow in order to find the location of the image.

Figure 24-7 Problem 41

As long as $d_O > f$, the image will be real in a concave, spherical mirror. In this case, the focal point is located between the mirror and the object.

REFLECT

An object placed outside the center of curvature of a concave, spherical mirror will result in an image smaller than the object; an object placed between the center of curvature and the focal point of a concave, spherical mirror will result in an image larger than the object. An object placed exactly at the center of curvature will result in an image that is the same size as the object.

24.43

SET UP

An object is placed in front of a concave mirror that has a radius of curvature of $r = 10.0$ cm. The object distance is $d_O = 8.00$ cm, and the focal length is $f = r/2 = +5.00$ cm. First, we can draw a ray trace diagram of the setup to determine the image distance and magnification of the image. A parallel ray from the tip of the object will reflect off of the mirror through the focal point. A ray from the tip of the object striking the center of the mirror is also easy to trace by applying the law of reflection. The image distance is the distance from the center of the mirror to the position of the image; the magnification is the ratio of the image height to the object height. An image located on the reflective side of the mirror is considered real. If the image is upside down relative to the object, it is said to be inverted. Secondly, we can apply the mirror equation, $\frac{1}{d_O} + \frac{1}{d_I} = \frac{1}{f}$, and its sign conventions to confirm the results from the ray trace diagram. A positive image distance is considered real, and a negative magnification corresponds to an inverted image.

SOLVE

Part a)

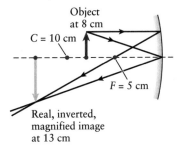

Figure 24-8 Problem 43

Part b)

Image distance:

$$\frac{1}{d_O} + \frac{1}{d_I} = \frac{1}{f}$$

$$\frac{1}{d_I} = \frac{1}{f} - \frac{1}{d_O} = \frac{1}{5.00 \text{ cm}} - \frac{1}{8.00 \text{ cm}} = \frac{3.00}{40.0 \text{ cm}}$$

$$d_I = \frac{40.0 \text{ cm}}{3.00} = \boxed{13.3 \text{ cm}}$$

Magnification:

$$m = -\frac{d_I}{d_O} = -\frac{13.3 \text{ cm}}{8.00 \text{ cm}} = \boxed{-1.66}$$

The image is $\boxed{\text{real}}$ because the image distance is positive. The image is $\boxed{\text{inverted}}$ because $m < 0$.

REFLECT

A ray trace diagram must be drawn to scale if you want to quantitatively measure distances and heights.

Get Help: Interactive Example – Spherical Mirror
P'Cast 24.1 – Images Made by a Concave Mirror

24.47

SET UP

The radius of curvature of a spherical concave mirror is 15 cm, which means its focal length is $f = +7.5$ cm. An object ($h_O = 20$ cm) is positioned at three different object distances: $d_O = +10$ cm, $d_O = +20$ cm, and $d_O = +100$ cm. We can use the mirror equation, $\frac{1}{d_O} + \frac{1}{d_I} = \frac{1}{f}$, and its sign conventions to calculate the image distance and the image height, as well as determine whether the image is real or virtual, upright or inverted. As a reminder, a positive image distance is considered real, and a negative image height corresponds to an inverted image.

SOLVE

Image distance in general:

$$\frac{1}{d_O} + \frac{1}{d_I} = \frac{1}{f}$$

$$\frac{1}{d_I} = \frac{1}{f} - \frac{1}{d_O} = \frac{d_O - f}{f d_O}$$

$$d_I = \frac{f d_O}{d_O - f}$$

Image height in general:

$$m = \frac{h_I}{h_O} = -\frac{d_I}{d_O}$$

$$h_I = -\frac{d_I}{d_O} h_O$$

Part a)

Image distance:

$$d_I = \frac{f d_O}{d_O - f} = \frac{(7.5 \text{ cm})(10 \text{ cm})}{(10 \text{ cm}) - (7.5 \text{ cm})} = \boxed{30 \text{ cm}}$$

The image is $\boxed{\text{real}}$ because $d_I > 0$.

Image height:

$$h_I = -\frac{d_I}{d_O} h_O = -\left(\frac{30 \text{ cm}}{10 \text{ cm}}\right)(20 \text{ cm}) = -60 \text{ cm}$$

The image is $\boxed{60 \text{ cm tall and inverted}}$ because $h_I < 0$.

Part b)

Image distance:

$$d_I = \frac{f d_O}{d_O - f} = \frac{(7.5 \text{ cm})(20 \text{ cm})}{(20 \text{ cm}) - (7.5 \text{ cm})} = 12 \text{ cm} \approx \boxed{10 \text{ cm (rounded to one significant figure)}}$$

The image is $\boxed{\text{real}}$ because $d_I > 0$.

Image height:

$$h_I = -\frac{d_I}{d_O} h_O = -\left(\frac{12 \text{ cm}}{20 \text{ cm}}\right)(20 \text{ cm}) = -12 \text{ cm} \approx -10 \text{ cm (rounded to one significant figure)}$$

The image is $\boxed{\text{about 10 cm tall and inverted}}$ because $h_I < 0$.

Part c)

Image distance:

$$d_I = \frac{f d_O}{d_O - f} = \frac{(7.5 \text{ cm})(100 \text{ cm})}{(100 \text{ cm}) - (7.5 \text{ cm})} = 8.1 \text{ cm} \approx \boxed{8 \text{ cm (rounded to one significant figure)}}$$

The image is $\boxed{\text{real}}$ because $d_I > 0$.

Chapter 24 Geometrical Optics

Image height:

$$h_I = -\frac{d_I}{d_O}h_O = -\left(\frac{8.1 \text{ cm}}{100 \text{ cm}}\right)(20 \text{ cm}) = -1.6 \text{ cm} \approx -2 \text{ cm (rounded to one significant figure)}$$

The image is about 2 cm tall and inverted because $h_I < 0$.

REFLECT

A real object placed at a position outside of the focal point of a spherical concave mirror will always give a real image:

$$d_I = \frac{fd_O}{d_O - f} > 0, \text{ if } (d_O - f) > 0$$

Get Help: Interactive Example – Spherical Mirror
P'Cast 24.1 – Images Made by a Concave Mirror

24.55

SET UP

The radius of curvature of a spherical convex mirror is 20 cm, which means its focal length is $f = -10$ cm. An object ($h_O = 10$ cm) is positioned at three different object distances: $d_O = +20$ cm, $d_O = +50$ cm, and $d_O = +100$ cm. We can use the mirror equation, $\frac{1}{d_O} + \frac{1}{d_I} = \frac{1}{f}$, and its sign conventions to calculate the image distance and the image height, as well as determine whether the image is real or virtual, upright or inverted. As a reminder, a positive image distance is considered real, and a negative image height corresponds to an inverted image.

SOLVE

Image distance in general:

$$\frac{1}{d_O} + \frac{1}{d_I} = \frac{1}{f}$$

$$\frac{1}{d_I} = \frac{1}{f} - \frac{1}{d_O} = \frac{d_O - f}{fd_O}$$

$$d_I = \frac{fd_O}{d_O - f}$$

Image height in general:

$$m = \frac{h_I}{h_O} = -\frac{d_I}{d_O}$$

$$h_I = -\frac{d_I}{d_O}h_O$$

Part a)

Image distance:

$$d_I = \frac{fd_O}{d_O - f} = \frac{(-10 \text{ cm})(20 \text{ cm})}{(20 \text{ cm}) - (-10 \text{ cm})} = -6.67 \text{ cm} \approx \boxed{-7 \text{ cm (rounded to one significant figure)}}$$

The image is virtual because $d_I < 0$.

Image height:

$$h_I = -\frac{d_I}{d_O}h_O = -\left(\frac{-6.67 \text{ cm}}{20 \text{ cm}}\right)(10 \text{ cm}) \approx 3 \text{ cm}$$

The image is $\boxed{\text{about 3 cm tall and upright}}$ because $h_I > 0$.

Part b)

Image distance:

$$d_I = \frac{fd_O}{d_O - f} = \frac{(-10 \text{ cm})(50 \text{ cm})}{(50 \text{ cm}) - (-10 \text{ cm})} = -8.33 \text{ cm} \approx \boxed{-8 \text{ cm (rounded to one significant figure)}}$$

The image is $\boxed{\text{virtual}}$ because $d_I < 0$.

Image height:

$$h_I = -\frac{d_I}{d_O}h_O = -\left(\frac{-8.33 \text{ cm}}{50 \text{ cm}}\right)(10 \text{ cm}) \approx 2 \text{ cm}$$

The image is $\boxed{\text{about 2 cm tall and upright}}$ because $h_I > 0$.

Part c)

Image distance:

$$d_I = \frac{fd_O}{d_O - f} = \frac{(-10 \text{ cm})(100 \text{ cm})}{(100 \text{ cm}) - (-10 \text{ cm})} = -9.09 \text{ cm} \approx \boxed{-9 \text{ cm (rounded to one significant figure)}}$$

The image is $\boxed{\text{virtual}}$ because $d_I < 0$.

Image height:

$$h_I = -\frac{d_I}{d_O}h_O = -\left(\frac{-9.09 \text{ cm}}{100 \text{ cm}}\right)(10 \text{ cm}) = 0.9 \text{ cm}$$

The image is $\boxed{\text{about 0.9 cm tall and upright}}$ because $h_I > 0$.

REFLECT

A real object placed *anywhere* in front of a spherical convex mirror will always give a virtual image, *i.e.*, $d_I = \frac{fd_O}{d_O - f} < 0$. The term $(d_O - f)$ will always be positive for a real object and a convex mirror.

Get Help: Interactive Example – Curved Mirror
P'Cast 24.2 – An Image Made by a Convex Mirror

24.61

SET UP

In order to prove that all images produced by a spherical convex mirror are virtual, we need to show mathematically that the image distance is negative regardless of our choice of d_O. We can rearrange the mirror equation, solve for the image distance, and apply the necessary sign conventions, mainly that a spherical convex mirror has a negative focal length.

84 Chapter 24 Geometrical Optics

SOLVE

$$\frac{1}{d_O} + \frac{1}{d_I} = \frac{1}{f}$$

$$\frac{1}{d_I} = \frac{1}{f} - \frac{1}{d_O} = \frac{d_O - f}{f d_O}$$

$$d_I = \frac{f d_O}{d_O - f}$$

The focal length of a spherical convex mirror is always negative, and the distance of the object from the mirror, d_O, is positive for all mirrors, which means the term $(d_O - f)$ will always be positive for a real object. Therefore,

$$d_I = \frac{f d_O}{d_O - f} < 0$$

REFLECT
A negative image distance means the image is virtual.

Get Help: Interactive Example – Curved Mirror
P'Cast 24.2 – An Image Made by a Convex Mirror

24.65

SET UP
A real image created by reflection in a spherical mirror appears in front of the mirrored surface. A real image created by refraction through a lens appears behind the lens.

SOLVE
$\boxed{\text{No}}$, a real image in a converging lens occurs on the opposite side of the lens from the object.

REFLECT
We expect light to be reflected by a mirror and transmitted by a lens.

Get Help: Picture It – Converging Lens

24.69

SET UP
The power of a converging lens is 5 diopters, which means its focal length is $f = \frac{1}{P} = \frac{1}{5 \text{ m}^{-1}} = 0.2 \text{ m} = +20 \text{ cm}$. An object ($h_O = 10$ cm) is positioned at four different object distances: $d_O = +5$ cm, $d_O = +10$ cm, $d_O = +20$ cm, and $d_O = +50$ cm. We can use the lens equation, $\frac{1}{d_O} + \frac{1}{d_I} = \frac{1}{f}$, and its sign conventions to calculate the image distance and the image height, as well as determine whether the image is real or virtual, upright or inverted. As a reminder, a positive image distance is considered real, and a negative image height corresponds to an inverted image. We can also draw a ray trace diagram of each situation to confirm our numerical results.

SOLVE
Image distance in general:

$$\frac{1}{d_O} + \frac{1}{d_I} = \frac{1}{f}$$

$$\frac{1}{d_I} = \frac{1}{f} - \frac{1}{d_O} = \frac{d_O - f}{f d_O}$$

$$d_I = \frac{f d_O}{d_O - f}$$

Image height in general:

$$m = \frac{h_I}{h_O} = -\frac{d_I}{d_O}$$

$$h_I = -\frac{d_I}{d_O} h_O$$

Part a)
Image distance:

$$d_I = \frac{f d_O}{d_O - f} = \frac{(20 \text{ cm})(5 \text{ cm})}{(5 \text{ cm}) - (20 \text{ cm})} = -6.7 \text{ cm} \approx \boxed{-7 \text{ cm (rounded to one significant figure)}}$$

The image is $\boxed{\text{virtual}}$ because $d_I < 0$.

Image height:

$$h_I = -\frac{d_I}{d_O} h_O = -\left(\frac{-6.7 \text{ cm}}{5 \text{ cm}}\right)(10 \text{ cm}) = 13 \text{ cm} \approx 10 \text{ cm (rounded to one significant figure)}$$

The image is $\boxed{\text{about 10 cm tall and upright}}$ because $h_I > 0$.

Ray trace diagram:

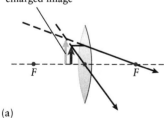

(a)
Figure 24-9 Problem 69

Part b)
Image distance:

$$d_I = \frac{f d_O}{d_O - f} = \frac{(20 \text{ cm})(10 \text{ cm})}{(10 \text{ cm}) - (20 \text{ cm})} = \boxed{-20 \text{ cm}}$$

The image is $\boxed{\text{virtual}}$ because $d_I < 0$.

Image height:

$$h_I = -\frac{d_I}{d_O}h_O = -\left(\frac{-20 \text{ cm}}{10 \text{ cm}}\right)(10 \text{ cm}) = 20 \text{ cm}$$

The image is $\boxed{20 \text{ cm tall and upright}}$ because $h_I > 0$.

Ray trace diagram:

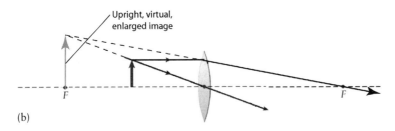

(b)

Figure 24-10 Problem 69

Part c)
Image distance:

$$d_I = \frac{fd_O}{d_O - f} = \frac{(20 \text{ cm})(20 \text{ cm})}{(20 \text{ cm}) - (20 \text{ cm})} = \infty$$

$\boxed{\text{No image is formed}}$.

Ray trace diagram:

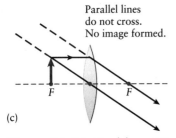

(c)

Figure 24-11 Problem 69

Part d)

Image distance:

$$d_I = \frac{fd_O}{d_O - f} = \frac{(20 \text{ cm})(50 \text{ cm})}{(50 \text{ cm}) - (20 \text{ cm})} = 33.3 \text{ cm} \approx \boxed{30 \text{ cm (rounded to one significant figure)}}$$

The image is $\boxed{\text{real}}$ because $d_I > 0$.

Image height:

$$h_I = -\frac{d_I}{d_O}h_O = -\left(\frac{33.3 \text{ cm}}{50 \text{ cm}}\right)(10 \text{ cm}) = -7 \text{ cm}$$

The image is $\boxed{\text{about 7 cm tall and inverted}}$ because $h_I < 0$.

Ray trace diagram:

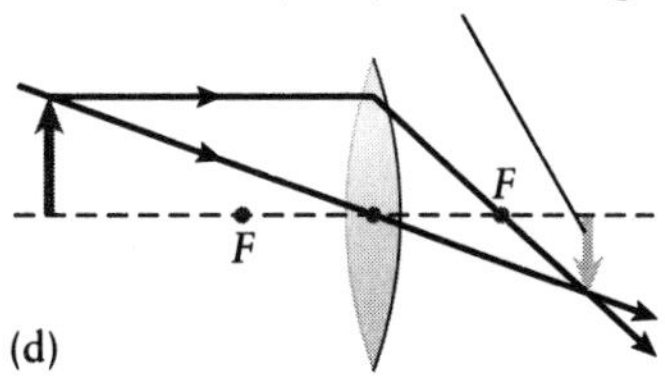

Inverted, real, reduced image

(d)

Figure 24-12 Problem 69

REFLECT
An object placed at the focal point of a lens will not create an image because the rays come out parallel and do not converge. This is the reverse situation of an object at infinity emitting parallel rays, which are converged at the focal point by a converging lens.

Get Help: Interactive Example – Focal Lens
P'Cast 24.3 – Calculating Focal Length for a Convex Lens
P'Cast 24.4 – Calculating Focal Length for a Concave Lens
P'Cast 24.5 – An Image Made by a Convex Lens

24.73

SET UP
A plano-concave, glass lens ($n = 1.60$) has a focal length of $f = -31.8$ cm. We can calculate the radius of curvature of the concave surface from the lensmaker's equation, $\frac{1}{f} = (n-1)\left(\frac{1}{R_1} - \frac{1}{R_2}\right)$. We'll assume that the front surface of the lens is planar, which means $R_1 = \infty$. For the lensmaker's equation, R_2 is positive for a concave surface and negative for a convex surface. Once we know the R_2 for a concave surface, we can reverse its sign in order to use the lensmaker's equation to calculate the focal length of a plano-convex, glass lens with the same radius of curvature.

SOLVE
Part a)

$$\frac{1}{f} = (n-1)\left(\frac{1}{R_1} - \frac{1}{R_2}\right)$$

$$\frac{1}{f(n-1)} = \frac{1}{R_1} - \frac{1}{R_2} = \frac{1}{\infty} - \frac{1}{R_2} = -\frac{1}{R_2}$$

$$R_2 = -f(n-1) = -(-31.8 \text{ cm})((1.60) - 1) = \boxed{19.1 \text{ cm} = 0.191 \text{ m}}$$

Part b)

$$\frac{1}{f} = (n-1)\left(\frac{1}{R_1} - \frac{1}{R_2}\right)$$

88   Chapter 24   Geometrical Optics

$$f = \frac{1}{(n-1)\left(\frac{1}{R_1} - \frac{1}{R_2}\right)} = \frac{1}{(n-1)\left(\frac{1}{\infty} - \frac{1}{R_2}\right)} = \frac{1}{(n-1)\left(-\frac{1}{R_2}\right)}$$

$$= \frac{-R_2}{n-1} = \frac{-(-19.1 \text{ cm})}{(1.60) - 1} = \boxed{+31.8 \text{ cm}}$$

REFLECT

It makes sense that changing the back surface of the lens from concave to convex should convert the diverging lens to a converging lens. Also, since we're not changing the magnitude of the radius of curvature of that surface, the magnitude of the focal length should not change.

Get Help: Interactive Example – Focal Lens
P'Cast 24.3 – Calculating Focal Length for a Convex Lens
P'Cast 24.4 – Calculating Focal Length for a Concave Lens
P'Cast 24.5 – An Image Made by a Convex Lens

General Problems

24.81

SET UP

A microscope that consists of an objective lens and an eyepiece separated by $L = 16$ cm has a total magnification of $m_{\text{total}} = 400$. The focal length of the objective is $f_o = 0.60$ cm. The magnification of each lens is determined by $m = \frac{-d_I}{d_O}$. The total magnification of a system of lenses is equal to the product of the magnifications due to each individual lens; in this case, $m_{\text{total}} = m_o m_e$. A microscope is constructed in such a way that a specimen located just beyond the focal plane of the objective lens creates an image near the focal plane of the eyepiece. Since the distance between the lenses is much larger than the focal length of the lenses, the magnification of the objective lens is equal to the distance separating the two lenses divided by the focal length of the objective lens. The eyepiece takes the image from the objective as its object and creates an image at the near point of your eye ($N = 25$ cm), which means the magnification due to the eyepiece is equal to N divided by the focal length of the eyepiece. By rearranging the expression for the total magnification, we can solve for the focal length of the eyepiece.

SOLVE

$$f_e = \frac{-LN}{f_o m_{\text{total}}} = \frac{-(16 \text{ cm})(25 \text{ cm})}{(0.60 \text{ cm})(400)} = \boxed{-2 \text{ cm (rounded to one significant figure)}}$$

REFLECT

In a compound microscope consisting of two lenses, the focal length of the objective lens is always smaller than the focal length of the eyepiece.

Get Help: Interactive Example – Focal Lens
P'Cast 24.3 – Calculating Focal Length for a Convex Lens
P'Cast 24.4 – Calculating Focal Length for a Concave Lens
P'Cast 24.5 – An Image Made by a Convex Lens

24.85

SET UP

A thin, diverging (concave) lens ($f_A = -45.0$ cm) has the same principal axis as a concave mirror with a radius of curvature of $+60.0$ cm. The focal length of a spherical mirror is equal to the radius of curvature divided in half, so $f_B = 30.0$ cm. An object is placed $d_{O,A} = 15.0$ cm in front of the lens. We will assume that the mirror and the object are on opposite sides of the lens, and the lens is small enough that light reflected by the mirror does not pass through the lens again. We can use the thin-lens equation to determine the location of the image produced by the lens. This image now acts as the object for the concave mirror, so we can use the mirror equation to calculate the location of the image after it was reflected by the mirror. If the image distance from the mirror calculation is positive, then the image is real; if it's negative, the image is virtual. Whether or not the image is upright or inverted can be found by calculating the total magnification due to both the lens and the mirror. A convex mirror of the same magnitude radius of curvature will have a focal length of $f_B = -30.0$ cm. We can follow the same process as the concave mirror to determine where the image is produced by this convex mirror.

SOLVE

Part a)

Image formed by the lens:

$$\frac{1}{d_{O,A}} + \frac{1}{d_{I,A}} = \frac{1}{f_A}$$

$$\frac{1}{d_{I,A}} = \frac{1}{f_A} - \frac{1}{d_{O,A}} = \frac{d_{O,A} - f_A}{f_A d_{O,A}}$$

$$d_{I,A} = \frac{f_A d_{O,A}}{d_{O,A} - f_A} = \frac{(-45.0 \text{ cm})(15.0 \text{ cm})}{(15.0 \text{ cm}) - (-45.0 \text{ cm})} = -11.25 \text{ cm}$$

Image formed by the mirror:

The image formed by the lens is located 11.25 cm in front of the lens, which means $d_{O,B} = (11.25 \text{ cm}) + (20.0 \text{ cm}) = 31.25$ cm.

$$\frac{1}{d_{O,B}} + \frac{1}{d_{I,B}} = \frac{1}{f_B}$$

$$\frac{1}{d_{I,B}} = \frac{1}{f_B} - \frac{1}{d_{O,B}} = \frac{d_{O,B} - f_B}{f_B d_{O,B}}$$

$$d_{I,B} = \frac{f_B d_{O,B}}{d_{O,B} - f_B} = \frac{(30.0 \text{ cm})(31.25 \text{ cm})}{(31.25 \text{ cm}) - (30.0 \text{ cm})} = 7.50 \times 10^2 \text{ cm}$$

The image formed by the mirror is located $\boxed{7.50 \times 10^2 \text{ cm in front of the mirror}}$ (or 7.30×10^2 cm in front of the lens).

Part b)

The final image is $\boxed{\text{real because } d_{I,B} > 0}$.

90   Chapter 24 Geometrical Optics

Magnification:

$$m_{\text{total}} = m_A m_B = \left(-\frac{d_{I,A}}{d_{O,A}}\right)\left(-\frac{d_{I,B}}{d_{O,B}}\right) = \left(-\frac{(-11.25 \text{ cm})}{(15.0 \text{ cm})}\right)\left(-\frac{7.50 \times 10^2 \text{ cm}}{31.25 \text{ cm}}\right) = -18.0$$

The image is $\boxed{\text{inverted because } m_{\text{total}} < 0}$.

Part c)

The only thing that changes is that $f_B = -30.0$ cm.

Therefore,

$$d_{I,B} = \frac{f_B d_{O,B}}{d_{O,B} - f_B} = \frac{(-30.0 \text{ cm})(31.25 \text{ cm})}{(31.25 \text{ cm}) - (-30.0 \text{ cm})} = -15.3 \text{ cm}$$

The image created by the mirror is $\boxed{\text{virtual because } d_{I,B} < 0}$.

Magnification:

$$m_{\text{total}} = m_A m_B = \left(-\frac{d_{I,A}}{d_{O,A}}\right)\left(-\frac{d_{I,B}}{d_{O,B}}\right) = \left(-\frac{(-11.25 \text{ cm})}{(15.0 \text{ cm})}\right)\left(-\frac{(-15.3 \text{ cm})}{(31.25 \text{ cm})}\right) = 0.367$$

The image is $\boxed{\text{upright because } m_{\text{total}} > 0}$.

REFLECT

If the lens were the same size as the mirror, the light reflected by the mirror will pass through the lens again, in the opposite direction as part (a). The real image created by the concave mirror will act as a virtual object for the lens in this case.

> **Get Help:** Interactive Example – Focal Lens
> P'Cast 24.3 – Calculating Focal Length for a Convex Lens
> P'Cast 24.4 – Calculating Focal Length for a Concave Lens
> P'Cast 24.5 – An Image Made by a Convex Lens

24.89

SET UP

The *tapetum lucidum* is a highly reflective membrane just behind the retina of the eyes of cats, which have a typical diameter of $d = 1.25$ cm. We can model this membrane as a concave spherical mirror with a radius of curvature of $r = d/2 = +0.625$ cm. We will assume that the light entering the cat's eye is traveling parallel to the principal axis of the lens. Parallel incoming light rays will be focused at the focal point; the focal length f for a concave spherical mirror is equal to $r/2$. The eye is filled with fluid with a refractive index of $n = 1.4$. Although this will affect the refraction of the light rays, it has no effect on reflection.

SOLVE

Part a)

$$f = \frac{r}{2} = \frac{0.625 \text{ cm}}{2} = 0.312 \text{ cm} = 3.12 \text{ mm}$$

The light will be focused $\boxed{3.12 \text{ mm in front of the retina}}$.

Part b) Reflection is not affected by the index of refraction, so the answer would be the same as in part (a).

REFLECT

We could have also calculated the image distance from the mirror equation:

$$\frac{1}{d_O} + \frac{1}{d_I} = \frac{1}{\infty} + \frac{1}{d_I} = \frac{1}{f}$$

$$d_I = f$$

Get Help: Interactive Example – Focal Lens
P'Cast 24.3 – Calculating Focal Length for a Convex Lens
P'Cast 24.4 – Calculating Focal Length for a Concave Lens
P'Cast 24.5 – An Image Made by a Convex Lens

24.93

SET UP

A specific person's near point was measured to be 15.0 cm and his far point is 2.75 m. These points for normal vision are 25.0 cm and infinity, respectively. It's not considered an issue if a person's near point is "too close"; his unaided eye can see things closer than an average person can, which is fine. Therefore, he wouldn't need to correct this issue and would only require lenses with a single focal length. The corrective contact lenses would need to take an object located at infinity and produce an image at the person's uncorrected far point. Contact lenses are located on a person's eye, so the image distance would be equal to the far point distance. Plugging all of this information with the correct sign conventions into the thin-lens equation will allow us to calculate the required focal length of the contact lenses. The power of the contact lenses is equal to the reciprocal of the focal length.

SOLVE

Part a) The person's far point is too close, so he needs a single focal length lens to correct this issue.

Part b)

$$\frac{1}{d_O} + \frac{1}{d_I} = \frac{1}{f}$$

$$\frac{1}{\infty} + \frac{1}{d_I} = \frac{1}{f}$$

$$f = d_I = \boxed{-2.75 \text{ m}}$$

Part c)

$$P = \frac{1}{f} = \frac{1}{-2.75 \text{ m}} = \boxed{-0.364 \text{ diopters}}$$

REFLECT

People who have trouble seeing things off in the distance are nearsighted because they can only see things near to them.

Get Help: Interactive Example – Focal Lens
P'Cast 24.3 – Calculating Focal Length for a Convex Lens
P'Cast 24.4 – Calculating Focal Length for a Concave Lens
P'Cast 24.5 – An Image Made by a Convex Lens

24.95

SET UP

We will assume that the focal length is known to 3 significant figures. A zoom lens for a digital camera is first zoomed such that the focal length is $f = 200$ mm $= 0.200$ m. The camera is focused on an object that is a distance $d_O = 15.0$ m from the lens. We can use the lens equation to calculate the distance between the lens and the photosensor array inside the camera, *i.e.*, the image distance d_I. The width of the image is essentially the same as the image height, h_I, which we can find from the definition of the magnification. The focal length of the zoom lens is then changed to $f = 18$ mm; since we already know what the image distance must be, we can calculate the closest an object can be for this case.

SOLVE

Part a)

$$\frac{1}{d_O} + \frac{1}{d_I} = \frac{1}{f}$$

$$\frac{1}{d_I} = \frac{1}{f} - \frac{1}{d_O} = \frac{d_O - f}{f d_O}$$

$$d_I = \frac{f d_O}{d_O - f} = \frac{(0.200 \text{ m})(15.0 \text{ m})}{(15.0 \text{ m}) - (0.200 \text{ m})} = \boxed{0.203 \text{ m} = 203 \text{ mm}}$$

Part b)

$$m = \frac{h_I}{h_O} = -\frac{d_I}{d_O}$$

$$h_I = -\frac{d_I}{d_O} h_O = -\left(\frac{0.203 \text{ m}}{15.0 \text{ m}}\right)(38 \text{ cm}) = \boxed{0.51 \text{ cm}}$$

Part c)

$$\frac{1}{d_O} + \frac{1}{d_I} = \frac{1}{f}$$

$$\frac{1}{d_O} = \frac{1}{f} - \frac{1}{d_I} = \frac{d_I - f}{f d_I}$$

$$d_O = \frac{f d_I}{d_I - f} = \frac{(18 \text{ mm})(52 \text{ mm})}{(52 \text{ mm}) - (18 \text{ mm})} = \boxed{28 \text{ mm}}$$

REFLECT

The image distance must remain constant since it is dictated by the dimensions and construction of the camera.

Get Help: Interactive Example – Focal Lens
P'Cast 24.3 – Calculating Focal Length for a Convex Lens
P'Cast 24.4 – Calculating Focal Length for a Concave Lens
P'Cast 24.5 – An Image Made by a Convex Lens

Chapter 25
Relativity

Conceptual Questions

25.3 A frame of reference or reference frame is a coordinate system with respect to which we will make observations or measurements. An inertial frame is one that moves at constant speed relative to another; that is, we refer to a frame of reference attached to a nonaccelerating object as an inertial frame.

Get Help: P'Cast 25.1 – Two Cars

25.7 If the object is moving relative to you, you measure its length by finding the difference between the coordinates of its end points at the same time.

Get Help: P'Cast 25.4 – A Flying Meter Stick I
P'Cast 25.5 – A Flying Meter Stick II

Multiple-Choice Questions

25.15 D (1.00c). The speed of light is constant in all reference frames.

25.19 A (The clock in Colorado runs slow). The effect of Earth's gravitational field decreases in magnitude as you move away from Earth.

Get Help: P'Cast 25.4 – A Flying Meter Stick I
P'Cast 25.5 – A Flying Meter Stick II

Estimation/Numerical Questions

25.25 About 5% of the world population.

Problems

25.29

SET UP

A frame of reference, S, is fixed on the surface of Earth with the x axis pointing toward the east, y axis pointing toward the north, and the z axis pointing up. A second frame of reference, S', is moving at a constant speed of $V = 4.00$ m/s to the east. We can use the Galilean transformation to mathematically describe the relationships between x and x', y and y', and z and z'. Once we have the expressions for x', y', and z', we can calculate their values given $t = 4.00$ s, $x = 2$ m, $y = 1$ m, and $z = 0$ m.

94   Chapter 25 Relativity

SOLVE

Part a)
$$x' = x - Vt$$
$$\boxed{x' = x - 4.00t} \text{ (SI units)}$$
$$\boxed{y' = y}$$
$$\boxed{z' = z}$$

Part b)
$$x' = x - Vt$$
$$x' = x - 4.00t = (2) - 4.00(4.00) = \boxed{-14 \text{ m}}$$
$$y' = y = \boxed{1 \text{ m}}$$
$$z' = z = \boxed{0 \text{ m}}$$

REFLECT
The second reference frame was moving toward the east ($+x$), so the sign of V should be positive.

25.35

SET UP
A float plane lands on a river. We will use a coordinate system where positive x points east and positive y points north. Let frame S be the frame of reference of an observer on the ground. Let frame S' be the frame of reference of the water, and frame S'' be the frame of reference of the wind. The relative speed of frames S'' and S is $V_{\text{wind to ground},y} = 20$ m/s (the velocity of the wind relative to the ground). The relative velocity of frames S' and S is $V_{\text{water to ground},y} = -5$ m/s. The velocity of the plane relative to the wind is $v''_{\text{plane},x} = 30$ m/s and $v''_{\text{plane},y} = 0$. Using the Galilean transformations, we can solve for the components of the velocity of the plane with respect to the water, $v'_{\text{plane},x}$ and $v'_{\text{plane},y}$. Note that because the wind and water do not move relative to the ground in the x direction, the East-West component of the airplane's velocity is the same in each reference frame. Thus, $v''_{\text{plane},x} = v'_{\text{plane},x} = 30$ m/s.

SOLVE

East-West component of plane velocity relative to water:
$$v''_{\text{plane},x} = v'_{\text{plane},x} = 30 \frac{\text{m}}{\text{s}}$$

North-South component of plane velocity relative to water:
Relating S and S'' frames:
$$v''_{\text{plane},y} = v_{\text{plane},y} - V_{\text{wind to ground},y}$$
$$v_{\text{plane},y} = v''_{\text{plane},y} + V_{\text{wind to ground},y}$$

Relating S and S' frames:
$$v'_{\text{plane},y} = v_{\text{plane},y} - V_{\text{water to ground},y}$$

Substituting equations:

$$v'_{\text{plane},y} = v''_{\text{plane},y} + V_{\text{wind to ground},y} - V_{\text{water to ground},y}$$

$$v'_{\text{plane},y} = 0 + 20\frac{\text{m}}{\text{s}} - \left(-5\frac{\text{m}}{\text{s}}\right) = \boxed{25\frac{\text{m}}{\text{s}}}$$

Magnitude:

$$v'_{\text{plane}} = \sqrt{(v'_{\text{plane},x})^2 + (v'_{\text{plane},y})^2} = \sqrt{\left(30\frac{\text{m}}{\text{s}}\right)^2 + \left(25\frac{\text{m}}{\text{s}}\right)^2} = \boxed{40\frac{\text{m}}{\text{s}}}$$

Direction:

$$\theta = \tan^{-1}\left(\frac{v'_{\text{plane},y}}{v'_{\text{plane},x}}\right) = \tan^{-1}\left(\frac{\left(25\frac{\text{m}}{\text{s}}\right)}{\left(30\frac{\text{m}}{\text{s}}\right)}\right) = \boxed{40°\text{ north of east}}$$

REFLECT

When relating frame S' to S, remember that the quantity V in the Galilean transformation is the speed of inertial frame S' relative to S.

Get Help: P'Cast 25.1 – Two Cars

25.43

SET UP

The average lifetime of muons traveling at a speed of $V = 0.98c$ was measured to be $\Delta t = 11$ μs. If the muons were at rest, then their average lifetime would be equal to the proper time $\Delta t_{\text{proper}} = \Delta t\sqrt{1 - \frac{V^2}{c^2}}$.

SOLVE

$$\Delta t_{\text{proper}} = \Delta t\sqrt{1 - \frac{V^2}{c^2}} = (11\ \mu\text{s})\sqrt{1 - \frac{(0.98c)^2}{c^2}} = (11\ \mu\text{s})\sqrt{1 - (0.98)^2} = \boxed{2.2\ \mu\text{s}}$$

REFLECT

The proper time is a time measurement made at rest with respect to the reference frame. We expect the lifetime measured for the moving muons to be longer than the ones at rest due to time dilation.

Get Help: P'Cast 25.2 – A Moving Clock

25.49

SET UP

A car, with a rest length of $L_{\text{rest}} = 3.20$ m, is moving in the x direction in the reference frame S. The reference frame S' moves at a speed of $V = 0.80c$ toward positive x. The length of the car according to observers in S' is equal to $L = L_{\text{rest}}\sqrt{1 - \frac{V^2}{c^2}}$.

SOLVE

$$L = L_{\text{rest}}\sqrt{1 - \frac{V^2}{c^2}} = (3.20\,\text{m})\sqrt{1 - \frac{(0.80c)^2}{c^2}}$$

$$= (3.20\,\text{m})\sqrt{1 - (0.80)^2} = \boxed{1.9\,\text{m}}$$

REFLECT

The observed length of the car in the moving reference frame should be smaller than the proper length of the car due to length contraction.

Get Help: P'Cast 25.2 – A Moving Clock

25.53

SET UP

We need to determine the speed of a pion that travels a distance of 100 m before it decays. We can define the distance traveled by the pion by imagining a ruler extending 100 m from Earth's surface to the initial location of the pion. Because the ruler is stationary relative to Earth, its length is $L_{\text{rest}} = 100$ m. The average lifetime, at rest, of a pion is $t_{\text{proper}} = 2.60 \times 10^{-8}$ s. The speed we're interested in is equal to the length of the ruler in the frame of the pion L divided by the lifetime of the pion t_{proper}. Since the pions are moving, the length of the ruler will appear to be contracted according to the equation: $L = L_{\text{rest}}\sqrt{1 - \frac{V^2}{c^2}}$. We can then solve for V to calculate the speed of the pions.

SOLVE

$$V = \frac{L}{t_{\text{proper}}} = \frac{L_{\text{rest}}\sqrt{1 - \frac{V^2}{c^2}}}{t_{\text{proper}}}$$

$$\frac{V}{\sqrt{1 - \frac{V^2}{c^2}}} = \frac{L_{\text{rest}}}{t_{\text{proper}}}$$

$$\left(\frac{V}{c\sqrt{1 - \frac{V^2}{c^2}}}\right)^2 = \left(\frac{L_{\text{rest}}}{t_{\text{proper}}c}\right)^2$$

$$\frac{\frac{V^2}{c^2}}{1 - \frac{V^2}{c^2}} = \left(\frac{L_{\text{rest}}}{t_{\text{proper}}c}\right)^2$$

$$\left(\frac{V}{c}\right)^2 = \frac{\left(\frac{L_{rest}}{t_{proper}c}\right)^2}{1+\left(\frac{L_{rest}}{t_{proper}c}\right)^2}$$

$$\frac{V}{c} = \sqrt{\frac{\left(\frac{L_{rest}}{t_{proper}c}\right)^2}{1+\left(\frac{L_{rest}}{t_{proper}c}\right)^2}} = \sqrt{\frac{\left(\frac{100\text{ m}}{(2.60\times 10^{-8}\text{ s})\left(3.00\times 10^8 \frac{\text{m}}{\text{s}}\right)}\right)^2}{1+\left(\frac{100\text{ m}}{(2.60\times 10^{-8}\text{ s})\left(3.00\times 10^8 \frac{\text{m}}{\text{s}}\right)}\right)^2}} = \boxed{0.997}$$

$$V = 0.997c = 0.997\left(3.00\times 10^8 \frac{\text{m}}{\text{s}}\right) = \boxed{2.99\times 10^8 \frac{\text{m}}{\text{s}}}$$

REFLECT

Moving at $0.997c$, the contracted length is:

$$L = L_{rest}\sqrt{1-\frac{V^2}{c^2}} = (100\text{ m})\sqrt{1-\frac{(0.997c)^2}{c^2}} \approx 8\text{ m}$$

In other words, in the pion frame of reference, the pion travels a distance of approximately 8 m.

Get Help: P'Cast 25.4 – A Flying Meter Stick I
P'Cast 25.5 – A Flying Meter Stick II

25.57

SET UP

A spaceship is traveling past Earth when it fires a rocket in the backward direction relative to the spaceship, and we want to know the velocity of the rocket relative to Earth. We will attach a stationary reference frame S to Earth and reference frame S' to the spaceship. The spaceship is traveling past Earth at a speed of $V = 0.92c$. The speed of the rocket relative to the spaceship is $v'_x = -0.75c$; the negative sign means the rocket is shot backward relative to the spaceship. We can rearrange the expression for the Lorentz transformation (Equation 25-17) in order to solve for the velocity of the rocket relative to Earth v_x.

SOLVE

$$v'_x = \frac{v_x - V}{1-\left(\frac{V}{c^2}v_x\right)}$$

$$v'_x\left(1-\left(\frac{V}{c^2}v_x\right)\right) = v_x - V$$

$$v'_x + V = v_x + \left(\frac{V}{c^2}v_x\right)v'_x$$

$$v_x = \frac{v'_x + V}{1 + \left(\frac{V}{c^2}v'_x\right)} = \frac{(-0.75c) + (0.92c)}{1 + \left(\frac{(0.92c)}{c^2}(-0.75c)\right)} = \frac{0.17c}{1 - 0.69} = \boxed{0.55c}$$

REFLECT

It makes sense that a rocket shot backward from a spaceship should be moving at a slower speed relative to Earth than the speed of the rocket relative to Earth.

Get Help: P'Cast 25.7 – Baseball for Superheroes

25.61

SET UP

An electron (mass $m = 9.11 \times 10^{-31}$ kg) travels at $v = 0.444c$. The magnitude of the Einsteinian momentum is $p = \gamma m v$, where $\gamma = \dfrac{1}{\sqrt{1 - \left(\frac{v}{c}\right)^2}}$. The total energy E of the electron is the sum of its Einsteinian kinetic energy K and the energy not associated with its motion, i.e., its rest energy, E_0. The rest energy is equal to $E_0 = mc^2$, and its total energy is equal to $E = \gamma mc^2$.

SOLVE

Relativistic gamma:

$$\gamma = \frac{1}{\sqrt{1 - \left(\frac{v}{c}\right)^2}} = \frac{1}{\sqrt{1 - \left(\frac{0.444c}{c}\right)^2}} = 1.116$$

Part a)

$$p = \gamma m v = \gamma m(0.444c)$$

$$= (1.116)(9.11 \times 10^{-31} \text{ kg})(0.444)\left(3.00 \times 10^8 \frac{\text{m}}{\text{s}}\right) = \boxed{1.35 \times 10^{-22} \frac{\text{kg} \cdot \text{m}}{\text{s}}}$$

Part b)

$$K + E_0 = E$$

$$K = E - E_0 = \gamma mc^2 - mc^2 = (\gamma - 1)mc^2 = ((1.116) - 1)mc^2$$

$$= (0.116)(9.11 \times 10^{-31} \text{ kg})\left(3.00 \times 10^8 \frac{\text{m}}{\text{s}}\right)^2 = \boxed{9.51 \times 10^{-15} \text{ J}}$$

Part c)

$$E_0 = mc^2 = (9.11 \times 10^{-31} \text{ kg})\left(3.00 \times 10^8 \frac{\text{m}}{\text{s}^2}\right)^2 = \boxed{8.20 \times 10^{-14} \text{ J}}$$

Part d)

$$E = \gamma mc^2 = (1.116)mc^2 = (1.116)(9.11 \times 10^{-31} \text{ kg})\left(3.00 \times 10^8 \frac{\text{m}}{\text{s}}\right)^2 = \boxed{9.15 \times 10^{-14} \text{ J}}$$

REFLECT

The work done on an object required to increase the speed approaches infinity since relativistic gamma approaches infinity as the object's speed approaches c.

Get Help: P'Cast 25.7 – Baseball for Superheroes

25.65

SET UP

A proton has a rest energy of $E_0 = mc^2 = 1.50 \times 10^{-10}$ J and a momentum of $p = \gamma mv = 1.07 \times 10^{-19} \frac{\text{kg} \cdot \text{m}}{\text{s}}$. To calculate the speed, we can relate the equation for momentum, $p = \gamma mv$, to the equation for rest energy, $E_0 = mc^2$.

SOLVE

$$p = \gamma mv$$

$$p^2 = \gamma^2 m^2 v^2$$

$$p^2 \left(\frac{1}{\gamma^2}\right) = m^2 v^2$$

$$p^2 \left(1 - \frac{v^2}{c^2}\right) = \left(\frac{E_0^2}{c^4}\right) v^2$$

$$p^2 - \frac{p^2 v^2}{c^2} = \left(\frac{E_0^2}{c^4}\right) v^2$$

$$p^2 = \left(\frac{E_0^2}{c^4} + \frac{p^2}{c^2}\right) v^2$$

$$p^2 = \left(\frac{E_0^2 + p^2 c^2}{c^4}\right) v^2$$

$$v^2 = \frac{p^2 c^4}{E_0^2 + p^2 c^2}$$

$$v = \left(\frac{pc}{\sqrt{E_0^2 + p^2 c^2}}\right) c$$

$$v = \left(\frac{\left(1.07 \times 10^{-19} \frac{\text{kg} \cdot \text{m}}{\text{s}}\right)\left(3.00 \times 10^8 \frac{\text{m}}{\text{s}}\right)}{\sqrt{(1.50 \times 10^{-10} \text{ J})^2 + \left(1.07 \times 10^{-19} \frac{\text{kg} \cdot \text{m}}{\text{s}}\right)^2 \left(3.00 \times 10^8 \frac{\text{m}}{\text{s}}\right)^2}}\right) c$$

$$\boxed{v = 0.209c}$$

100 Chapter 25 Relativity

REFLECT
We couldn't have simply divided the momentum by γm to solve for v because γ is also a function of v.

25.67

SET UP
An elevator near Earth's surface is accelerating downward at 18.0 m/s². We can use the general theory of relativity to determine the free-fall acceleration an observer inside the elevator would experience.

SOLVE

$$a_y = \left(-9.80\frac{m}{s^2}\right) - \left(-18.0\frac{m}{s^2}\right) = 8.20\frac{m}{s^2}$$

The observer in the elevator would measure the free-fall acceleration as $\boxed{8.20 \text{ m/s}^2 \text{ in the upward direction}}$.

REFLECT
We can do our own thought experiment to confirm our answer. The observer inside the elevator throws a ball horizontally. If the elevator car were stationary in free space, the ball would go in a straight line until it hit the wall. If the elevator car were accelerating downward, the ball would hit the wall at a location *higher* than the previous case because the entire elevator would have shifted downward due to its acceleration.

25.69

SET UP
Observers in a reference frame S see one explosion at $x_1 = 580$ m and then a second explosion $\Delta t = 4.5$ μs later at $x_2 = 1500$ m. Reference frame S' is moving along the positive x axis at a speed v. We can calculate v from the data measured by those in S. Observers in S' see the explosions occur at the same point in space. Since the observers in S' see the explosions occur at the same point in space, we know that the proper time is the time separation measured in S'. The time between explosions as measured in S' is related to the time interval measured in S through relativistic gamma, $\Delta t = \dfrac{\Delta t_{\text{proper}}}{\sqrt{1 - \dfrac{V^2}{c^2}}}$.

SOLVE
Speed

$$v = \frac{x_2 - x_1}{\Delta t} = \frac{(1500 \text{ m}) - (580 \text{ m})}{4.5 \times 10^{-6} \text{ s}} = 2.044 \times 10^8 \frac{m}{s}$$

Time separation

$$\Delta t = \frac{\Delta t_{\text{proper}}}{\sqrt{1 - \dfrac{V^2}{c^2}}}$$

$$\Delta t_{\text{proper}} = \Delta t \sqrt{1 - \left(\frac{v}{c}\right)^2} = (4.5 \text{ μs}) \sqrt{1 - \left(\frac{\left(2.044 \times 10^8 \frac{\text{m}}{\text{s}}\right)}{\left(3.00 \times 10^8 \frac{\text{m}}{\text{s}}\right)}\right)^2} = \boxed{3.3 \text{ μs}}$$

REFLECT

The proper time interval should be smaller than the time interval observed by those in S due to the effects of time dilation.

25.75

SET UP

A jet plane flies at $V = 300$ m/s relative to an observer on the ground. There is a clock aboard the plane, as well as one on the ground; the two were synchronized at the start. We want to know the distance (as measured by the observer on the ground) the plane must fly such that the clock on the plane is 10 s behind the clock on the ground. This time difference is the difference between the dilated time experienced on the plane and the proper time experienced on the ground. From this, we can calculate the proper time; the distance the plane must fly is equal to the plane's speed relative to the ground multiplied by the proper time. To simplify the calculation, it will be helpful to use the approximation for $(1-x)^{-\frac{1}{2}}$ when x is much smaller than 1: $(1-x)^{-\frac{1}{2}} \approx 1 + \frac{1}{2}x$.

SOLVE

Time observed on the ground:

$$t - t_{\text{proper}} = \frac{t_{\text{proper}}}{\sqrt{1 - \frac{V^2}{c^2}}} - t_{\text{proper}} \approx t_{\text{proper}}\left(1 + \frac{1}{2}\left(\frac{V^2}{c^2}\right) - 1\right) = 10 \text{ s}$$

$$t_{\text{proper}} \approx \frac{(20 \text{ s})c^2}{V^2}$$

Distance of the flight as measured by an observer on the ground:

$$V = \frac{\Delta x}{t_{\text{proper}}}$$

$$\Delta x = V t_{\text{proper}} \approx V\left(\frac{(20 \text{ s})c^2}{V^2}\right) \approx \frac{(20 \text{ s})c^2}{V}$$

$$\Delta x \approx \frac{(20 \text{ s})\left(3.00 \times 10^8 \frac{\text{m}}{\text{s}}\right)^2}{\left(300 \frac{\text{m}}{\text{s}}\right)} \approx \boxed{6 \times 10^{15} \text{ m}}$$

102 Chapter 25 Relativity

REFLECT

This is approximately 0.6 ly. For comparison, the distance between Earth and the Sun is about 1.5×10^{11} m; the distance the jet has to travel is 40,000 times larger than that.

25.79

SET UP

In one month, a household uses 411 kWh of electrical energy and 201 therms of gas heating (1.0 therm = 29.3 kWh). After converting these values into joules, we can use $E_0 = mc^2$ to calculate the mass necessary to meet the monthly energy needs for this house.

SOLVE

Conversions:

$$411 \text{ kWh} \times \frac{3600 \text{ s}}{1 \text{ h}} \times \frac{1000 \text{ W}}{1 \text{ kW}} = 1.480 \times 10^9 \text{ J}$$

$$201 \text{ therms} \times \frac{29.3 \text{ kWh}}{1.0 \text{ therm}} \times \frac{3600 \text{ s}}{1 \text{ h}} \times \frac{1000 \text{ W}}{1 \text{ kW}} = 2.120 \times 10^{10} \text{ J}$$

Rest energy:

$$E_0 = (2.120 \times 10^{10} \text{ J}) + (1.480 \times 10^9 \text{ J}) = 2.268 \times 10^{10} \text{ J}$$

Mass:

$$E_0 = mc^2$$

$$m = \frac{E_0}{c^2} = \frac{2.268 \times 10^{10} \text{ J}}{\left(3.00 \times 10^8 \frac{\text{m}}{\text{s}}\right)^2} = 2.52 \times 10^{-7} \text{ kg} \times \frac{10^6 \text{ mg}}{1 \text{ kg}} = \boxed{0.252 \text{ mg}}$$

REFLECT

This would be the mass necessary if we were harvesting the energy directly from the atoms that make up the molecules.

25.87

SET UP

A 25-year-old captain pilots a spaceship to a planet that is $\Delta x = 42$ ly from Earth. The captain needs to be no more than 60 years old when she arrives at the planet, which means the proper time interval, as observed by the spaceship, is $\Delta t_{\text{proper}} = 35$ y. The speed of the spaceship relative to Earth v is equal to Δx divided by the time interval as observed by Earth Δt, where $\Delta t = \dfrac{\Delta t_{\text{proper}}}{\sqrt{1 - \dfrac{v^2}{c^2}}}$. In order to make the math easier, we'll use the fact that 1 ly = (1 y)c. The captain sends a radio signal back to Earth as soon as it reaches the planet. The total time after the launch when the signal arrives is equal to the time it takes the ship to reach the planet plus the time it takes the signal to travel back to Earth. We can relate these times to the distance between Earth and the planet as well as the speeds of the spaceship and the speed of light.

SOLVE

Part a)

$$\Delta t = \frac{\Delta t_{proper}}{\sqrt{1 - \frac{V^2}{c^2}}}$$

$$v = \frac{\Delta x}{\Delta t} = \frac{\Delta x}{\Delta t_{proper}}\sqrt{1 - \left(\frac{v}{c}\right)^2}$$

$$v^2 = \left(\frac{\Delta x}{\Delta t_{proper}}\right)^2 - \left(\frac{\Delta x}{\Delta t_{proper}}\right)^2\left(\frac{v}{c}\right)^2$$

$$v = \frac{\Delta x}{\Delta t_{proper}\sqrt{1 + \left(\frac{\Delta x}{c\Delta t_{proper}}\right)^2}} = \frac{c\Delta x}{\sqrt{(c\Delta t_{proper})^2 + (\Delta x)^2}} = \frac{c(42\ \text{ly})}{\sqrt{(c(35\ \text{y}))^2 + (42\ \text{ly})^2}}$$

$$= \frac{c((42\ \text{y})c)}{\sqrt{(c(35\ \text{y}))^2 + ((42\ \text{y})c)^2}} = \frac{c(42\ \text{y})}{\sqrt{(35\ \text{y})^2 + (42\ \text{y})^2}} = \boxed{0.77c}$$

Part b)

$$\Delta t_{total} = \Delta t_{spaceship} + \Delta t_{signal} = \frac{\Delta x}{v} + \frac{\Delta x}{c} = \Delta x\left(\frac{1}{v} + \frac{1}{c}\right) = \Delta x\left(\frac{1}{0.77c} + \frac{1}{c}\right) = \frac{\Delta x}{c}\left(\frac{1}{0.77} + 1\right)$$

$$= \frac{(42\ \text{ly})}{c}(2.30) = \frac{(42\ \text{y})c}{c}(2.30) = \boxed{97\ \text{y}}$$

REFLECT

A distance of 42 ly is around 4×10^{17} m!

Get Help: P'Cast 25.9 – Converting Mass to Energy in the Sun

Chapter 26
Quantum Physics and Atomic Structure

Conceptual Questions

26.5 The shortest wavelength is the wavelength that is emitted when a free electron is captured into the lowest energy state. In this scenario, we consider an electron falling from the mth energy level at infinity to the lowest energy level, $n = 1$.

$$\frac{1}{\lambda} = R_H\left(\frac{1}{n^2} - \frac{1}{m^2}\right)$$

$$\text{As } m \to \infty, \frac{1}{m^2} \to 0$$

$$\frac{1}{\lambda} = R_H\left(\frac{1}{n^2} - 0\right) = R_H\left(\frac{1}{(1)^2}\right) = R_H$$

$$\lambda = \frac{1}{R_H} = \frac{1}{(1.097 \times 10^7 \text{m}^{-1})} = \boxed{91.16 \text{ nm}}$$

Get Help: P'Cast 26.4 – Photon Possibilities

26.13 Assuming nonrelativistic speeds, we can express a particle's kinetic energy K in terms of its de Broglie wavelength λ as follows:

$$K = \frac{1}{2}mv^2 = \frac{p^2}{2m} = \frac{\left(\frac{h}{\lambda}\right)^2}{2m} = \frac{h^2}{2m\lambda^2}$$

We can use the relationship between the kinetic energies of the electron, K_e, and the proton, K_p, to relate the masses of the particles to their de Broglie wavelengths:

$$K_e = K_p$$

$$\frac{h^2}{2m_e\lambda_e^2} = \frac{h^2}{2m_p\lambda_p^2}$$

$$\frac{m_p}{m_e} = \frac{\lambda_e^2}{\lambda_p^2}$$

Since $m_p \gg m_e$, we can conclude that $\lambda_e \gg \lambda_p$. The electron has a longer de Broglie wavelength.

Get Help: P'Cast 26.3 – Finding the Wavelength of a Room-Temperature Neutron

26.17 Because the mass of a macroscopic object is so large, the de Broglie wavelength is too small to observe. The wavelength of everyday objects is many orders of magnitude less than the radius of an atom. This is far too small for us to observe.

Get Help: P'Cast 26.3 – Finding the Wavelength of a Room-Temperature Neutron

Multiple-Choice Questions

26.21 A (a photon of ultraviolet radiation). Out of all of the choices, ultraviolet light has the highest frequency and, thus, the highest energy.

26.25 C (decreases). The increase in wavelength in the Compton effect as a function of wavelength is $\Delta\lambda \propto (1 - \cos(\theta))$. The energy is inversely proportional to the wavelength.

> **Get Help:** Picture It – Compton Scattering
> P'Cast 26.2 – Compton Scattering

Estimation/Numerical Analysis

26.29 Assume the body's surface area is about 2 m, the body temperature is 307 K, and a skin emissivity of almost 1. The rate of energy flow is:

$$P = e\sigma A T^4 \approx (1)(5.6704 \times 10^{-8} \text{ W} \cdot \text{m}^{-2} \cdot \text{K}^{-4})(2 \text{ m})(307 \text{ K}) \approx 1000 \text{ W}$$

26.33 An electron must travel at 7×10^6 m/s for its de Broglie wavelength to be about the size of an atom (about 10^{-10} m).

$$\lambda = \frac{h}{p} = \frac{h}{m_e v}$$

$$v = \frac{h}{m_e \lambda}$$

$$v = \frac{6.63 \times 10^{-34} \text{ J} \cdot \text{s}}{(9.11 \times 10^{-31} \text{ kg})(10^{-10} \text{ m})}$$

$$v \approx 7 \times 10^6 \frac{\text{m}}{\text{s}}$$

> **Get Help:** P'Cast 26.3 – Finding the Wavelength of a Room-Temperature Neutron

Problems

26.39

SET UP

A blackbody at $T = 300$ K radiates heat into its immediate surroundings. Wien's displacement law, $\lambda_{max} = \frac{0.290 \text{ K} \cdot \text{cm}}{T}$, tells us the wavelength at which the maximum intensity is emitted.

SOLVE

$$\lambda_{max} = \frac{0.290 \text{ K} \cdot \text{cm}}{300 \text{ K}} = \boxed{9.67 \times 10^{-4} \text{ cm} \approx 10{,}000 \text{ nm}}$$

REFLECT

This corresponds to infrared light, which makes sense for a blackbody at a relatively low temperature.

26.43

SET UP

We are asked to calculate the range of frequencies and energies in the visible spectrum of light, which is approximately from $\lambda_{\text{violet}} = 380 \times 10^{-9}$ m to $\lambda_{\text{red}} = 750 \times 10^{-9}$ m. The frequency of a photon is inversely related to its wavelength, $f = \dfrac{c}{\lambda}$; the energy is directly proportional to the frequency, $E = hf$.

SOLVE

Frequencies:

$$f_{\text{violet}} = \frac{c}{\lambda_{\text{violet}}} = \frac{\left(3.00 \times 10^8 \dfrac{\text{m}}{\text{s}}\right)}{380 \times 10^{-9} \text{ m}}$$

$$= 7.89 \times 10^{14} \text{ Hz} \approx \boxed{7.9 \times 10^{14} \text{ Hz (rounded to 2 significant figures)}}$$

$$f_{\text{red}} = \frac{c}{\lambda_{\text{red}}} = \frac{\left(3.00 \times 10^8 \dfrac{\text{m}}{\text{s}}\right)}{750 \times 10^{-9} \text{ m}}$$

$$= 4.00 \times 10^{14} \text{ Hz} \approx \boxed{4.0 \times 10^{14} \text{ Hz (rounded to 2 significant figures)}}$$

Energies:

$$E_{\text{violet}} = hf_{\text{violet}} = (6.63 \times 10^{-34} \text{ J} \cdot \text{s})(7.89 \times 10^{14} \text{ Hz}) = \boxed{5.2 \times 10^{-19} \text{ J}}$$

$$E_{\text{red}} = hf_{\text{red}} = (6.63 \times 10^{-34} \text{ J} \cdot \text{s})(4.00 \times 10^{14} \text{ Hz}) = \boxed{2.7 \times 10^{-19} \text{ J}}$$

REFLECT

Keep in mind that the wavelength is inversely proportional to the energy; red light has a longer wavelength than violet light, but is less energetic.

26.49

SET UP

Light ($\lambda = 195 \times 10^{-9}$ m) strikes a metal surface and photoelectrons are produced moving with a maximum speed of $v_{\text{max}} = 0.004c$. The expression relating the maximum kinetic energy of the photoelectrons to the work function is $K_{\text{max}} = hf - \Phi_0$. The threshold wavelength for the metal corresponds to the wavelength absorbed when the kinetic energy of the photoelectrons is equal to zero.

SOLVE

Part a)

$$K_{\text{max}} = hf - \Phi_0$$

$$\Phi_0 = hf - K_{\text{max}} = hf - \frac{1}{2}m_e v_{\text{max}}^2 = \frac{hc}{\lambda} - \frac{1}{2}m_e(0.004c)^2$$

$$= \frac{(6.63 \times 10^{-34} \text{ J} \cdot \text{s})\left(3.00 \times 10^8 \frac{\text{m}}{\text{s}}\right)}{195 \times 10^{-9} \text{ m}} - \frac{1}{2}(9.11 \times 10^{-31} \text{ kg})(0.004)^2\left(3.00 \times 10^8 \frac{\text{m}}{\text{s}}\right)^2$$

$$= \boxed{3.64 \times 10^{-19} \text{ J}}$$

Part b)

$$K_{max} = 0 = hf_{min} - \Phi_0 = \frac{hc}{\lambda_{max}} - \Phi_0$$

$$\lambda_{max} = \frac{hc}{\Phi_0} = \frac{(6.63 \times 10^{-34} \text{ J} \cdot \text{s})\left(3.00 \times 10^8 \frac{\text{m}}{\text{s}}\right)}{3.64 \times 10^{-19} \text{ J}} = \boxed{5.46 \times 10^{-7} \text{ m} = 546 \text{ nm}}$$

REFLECT

We can't know the exact type of metal without more information and a more precise measurement.

Get Help: Interactive Example – Photoelectric Effect I
P'Cast 26.1 – The Photoelectric Effect with Cesium

26.53

SET UP

The momentum of a photon is equal to the energy of the photon divided by the speed of light. By using the expression for the energy in terms of the wavelength of the light, we can easily calculate the momenta of photons with wavelengths of 550×10^{-9} m and 0.0711×10^{-9} m.

SOLVE

Relating momentum to wavelength:

$$p = \frac{E}{c} = \frac{h}{\lambda}$$

Part a)

$$p = \frac{h}{\lambda} = \frac{6.63 \times 10^{-34} \text{ J} \cdot \text{s}}{550 \times 10^{-9} \text{ m}} = \boxed{1.2 \times 10^{-27} \frac{\text{kg} \cdot \text{m}}{\text{s}}}$$

Part b)

$$p = \frac{h}{\lambda} = \frac{6.63 \times 10^{-34} \text{ J} \cdot \text{s}}{0.0711 \times 10^{-9} \text{ m}} = \boxed{9.32 \times 10^{-24} \frac{\text{kg} \cdot \text{m}}{\text{s}}}$$

REFLECT

We can also use the fact that $hc = 1240$ eV · nm to represent the momentum in units of eV/c. Our answers in these units would be 2.3 eV/c and 17,400 eV/c, respectively.

Get Help: Picture It – Compton Scattering
P'Cast 26.2 – Compton Scattering

26.57

SET UP

Photons with a wavelength of $\lambda_i = 0.14000$ nm are scattered at varying angles—0.00°, 30.0°, 45.0°, 60.0°, 90.0°, and 180°—off of carbon atoms. The shift in wavelength of the Compton scattered photons is given by $\Delta\lambda = \dfrac{h}{m_e c}(1 - \cos(\theta))$, where $\dfrac{h}{m_e c} = \lambda_C = 0.00243$ nm; the wavelength of the scattered photons can be calculated from this relationship. The kinetic energy of the scattered electrons is equal to the difference between the energy of the initial photons and the energy of the scattered photons.

SOLVE

Wavelength of scattered photon:

$$\Delta\lambda = \lambda_f - \lambda_i = \lambda_C(1 - \cos(\theta))$$

$$\lambda_f = \lambda_i + \lambda_C(1 - \cos(\theta))$$

The kinetic energy K of scattered electrons is the difference in the initial photon energy, E_i, and the final photon energy, E_f:

$$\frac{E}{c} = \frac{h}{\lambda}$$

$$K = E_i - E_f = \frac{hc}{\lambda_i} - \frac{hc}{\lambda_f} = hc\left(\frac{1}{\lambda_i} - \frac{1}{\lambda_f}\right)$$

The quantity hc can be expressed in units of J · nm:

$$hc = (6.63 \times 10^{-34}\,\text{J}\cdot\text{s})\left(3.00 \times 10^8\,\frac{\text{m}}{\text{s}}\right)\left(\frac{10^9\,\text{nm}}{1\,\text{m}}\right) = 1.989 \times 10^{-16}\,\text{J}\cdot\text{nm}$$

Part a)

$$\lambda_f = \lambda_i + \lambda_C(1 - \cos(\theta)) = (0.14000\,\text{nm}) + (0.00243\,\text{nm})(1 - \cos(0.00°)) = \boxed{0.14000\,\text{nm}}$$

$$K = hc\left(\frac{1}{\lambda_i} - \frac{1}{\lambda_f}\right) = (1.989 \times 10^{-16}\,\text{J}\cdot\text{nm})\left(\frac{1}{0.14000\,\text{nm}} - \frac{1}{0.14000\,\text{nm}}\right) = \boxed{0}$$

Part b)

$$\lambda_f = \lambda_i + \lambda_C(1 - \cos(\theta)) = (0.14000\,\text{nm}) + (0.00243\,\text{nm})(1 - \cos(30.0°)) = \boxed{0.14033\,\text{nm}}$$

$$K = hc\left(\frac{1}{\lambda_i} - \frac{1}{\lambda_f}\right) = (1.989 \times 10^{-16}\,\text{J}\cdot\text{nm})\left(\frac{1}{0.14000\,\text{nm}} - \frac{1}{0.14033\,\text{nm}}\right) = \boxed{3.34 \times 10^{-18}\,\text{J}}$$

Part c)

$$\lambda_f = \lambda_i + \lambda_C(1 - \cos(\theta)) = (0.14000\,\text{nm}) + (0.00243\,\text{nm})(1 - \cos(45.0°)) = \boxed{0.14071\,\text{nm}}$$

$$K = hc\left(\frac{1}{\lambda_i} - \frac{1}{\lambda_f}\right) = (1.989 \times 10^{-16}\,\text{J}\cdot\text{nm})\left(\frac{1}{0.14000\,\text{nm}} - \frac{1}{0.14071\,\text{nm}}\right) = \boxed{7.17 \times 10^{-18}\,\text{J}}$$

Part d)

$\lambda_f = \lambda_i + \lambda_C(1 - \cos(\theta)) = (0.14000 \text{ nm}) + (0.00243 \text{ nm})(1 - \cos(60.0°)) = \boxed{0.14122 \text{ nm}}$

$K = hc\left(\dfrac{1}{\lambda_i} - \dfrac{1}{\lambda_f}\right) = (1.989 \times 10^{-16} \text{ J} \cdot \text{nm})\left(\dfrac{1}{0.14000 \text{ nm}} - \dfrac{1}{0.14122 \text{ nm}}\right) = \boxed{1.23 \times 10^{-17} \text{ J}}$

Part e)

$\lambda_f = \lambda_i + \lambda_C(1 - \cos(\theta)) = (0.14000 \text{ nm}) + (0.00243 \text{ nm})(1 - \cos(90.0°)) = \boxed{0.14243 \text{ nm}}$

$K = hc\left(\dfrac{1}{\lambda_i} - \dfrac{1}{\lambda_f}\right) = (1.989 \times 10^{-16} \text{ J} \cdot \text{nm})\left(\dfrac{1}{0.14000 \text{ nm}} - \dfrac{1}{0.14243 \text{ nm}}\right) = \boxed{2.42 \times 10^{-17} \text{ J}}$

Part f)

$\lambda_f = \lambda_i + \lambda_C(1 - \cos(\theta)) = (0.14000 \text{ nm}) + (0.00243 \text{ nm})(1 - \cos(180.0°)) = \boxed{0.14486 \text{ nm}}$

$K = hc\left(\dfrac{1}{\lambda_i} - \dfrac{1}{\lambda_f}\right) = (1.989 \times 10^{-16} \text{ J} \cdot \text{nm})\left(\dfrac{1}{0.14000 \text{ nm}} - \dfrac{1}{0.14486 \text{ nm}}\right) = \boxed{4.77 \times 10^{-17} \text{ J}}$

REFLECT

Since the wavelength of a photon is inversely proportional to its energy, the maximum wavelength of the scattered photon and the maximum kinetic energy of the scattered electrons both occur when the scattering angle is 180°.

Get Help: Picture It – Compton Scattering
P'Cast 26.2 – Compton Scattering

26.63

SET UP

An electron ($m_e = 9.11 \times 10^{-31}$ kg) has a speed of $v = 0.00730c$. Its de Broglie wavelength is equal to Planck's constant divided by its momentum.

SOLVE

$\lambda = \dfrac{h}{p} = \dfrac{h}{m_e v} = \dfrac{6.63 \times 10^{-34} \text{ J} \cdot \text{s}}{(9.11 \times 10^{-31} \text{ kg})(0.00730)\left(3.00 \times 10^8 \dfrac{\text{m}}{\text{s}}\right)} = \boxed{3.32 \times 10^{-10} \text{ m}}$

REFLECT

A distance of 3 angstroms (1 angstrom = 0.1 nm) is a little larger than a bond length in a molecule; this is why electrons are commonly used in microscopy to probe molecular structure. The exact de Broglie wavelength of the electron can be tuned by changing its speed.

Get Help: P'Cast 26.3 – Finding the Wavelength of a Room-Temperature Neutron

26.67

SET UP

We are given the kinetic energies of six different electrons, 1.60×10^{-19} J, 1.60×10^{-18} J, 1.60×10^{-17} J, 1.60×10^{-16} J, 1.60×10^{-13} J, 1.60×10^{-10} J, and are asked to find the de Broglie wavelength, $\lambda = \dfrac{h}{p}$, of each one. First, we need to determine whether we need to use the expression for the classical or relativistic momentum; relativistic effects should be included if the speed of the particle is about 10% the speed of light. If the speed we calculate from the kinetic energy is less than $0.1c$, then we can use the classical expressions for the kinetic energy $\left(K = \frac{1}{2}m_e v^2\right)$ and momentum ($p = m_e v$). If the speed is larger than $0.1c$, we need to use the relativistic expressions for the kinetic energy, $K = (\gamma - 1)m_e c^2$, and momentum, $p = \gamma m_e v$.

SOLVE

Part a)

Speed:

$$K = \frac{1}{2}m_e v^2$$

$$v = \sqrt{\frac{2K}{m_e}} = \sqrt{\frac{2(1.60 \times 10^{-19} \text{ J})}{9.11 \times 10^{-31} \text{ kg}}} = 5.927 \times 10^5 \frac{\text{m}}{\text{s}}$$

Ratio of v to c:

$$\frac{v}{c} = \frac{\left(5.927 \times 10^5 \frac{\text{m}}{\text{s}}\right)}{\left(3.00 \times 10^8 \frac{\text{m}}{\text{s}}\right)} = 1.98 \times 10^{-3}, \text{ so we can safely ignore relativistic effects.}$$

de Broglie wavelength:

$$\lambda = \frac{h}{p} = \frac{h}{m_e v} = \frac{6.63 \times 10^{-34} \text{ J} \cdot \text{s}}{(9.11 \times 10^{-31} \text{ kg})\left(5.927 \times 10^5 \frac{\text{m}}{\text{s}}\right)} = \boxed{1.23 \times 10^{-9} \text{ m}}$$

Part b)

Speed:

$$K = \frac{1}{2}m_e v^2$$

$$v = \sqrt{\frac{2K}{m_e}} = \sqrt{\frac{2(1.60 \times 10^{-18} \text{ J})}{9.11 \times 10^{-31} \text{ kg}}} = 1.874 \times 10^6 \frac{\text{m}}{\text{s}}$$

Ratio of v to c:

$$\frac{v}{c} = \frac{\left(1.874 \times 10^6 \frac{\text{m}}{\text{s}}\right)}{\left(3.00 \times 10^8 \frac{\text{m}}{\text{s}}\right)} = 6.25 \times 10^{-3}, \text{ so we can safely ignore relativistic effects.}$$

de Broglie wavelength:

$$\lambda = \frac{h}{p} = \frac{h}{m_e v} = \frac{6.63 \times 10^{-34} \text{ J} \cdot \text{s}}{(9.11 \times 10^{-31} \text{ kg})\left(1.874 \times 10^6 \frac{\text{m}}{\text{s}}\right)} = \boxed{3.88 \times 10^{-10} \text{ m}}$$

Part c)

Speed:

$$K = \frac{1}{2}m_e v^2$$

$$v = \sqrt{\frac{2K}{m_e}} = \sqrt{\frac{2(1.60 \times 10^{-17}\,\text{J})}{9.11 \times 10^{-31}\,\text{kg}}} = 5.927 \times 10^6 \frac{\text{m}}{\text{s}}$$

Ratio of v to c:

$$\frac{v}{c} = \frac{\left(5.927 \times 10^6 \frac{\text{m}}{\text{s}}\right)}{\left(3.00 \times 10^8 \frac{\text{m}}{\text{s}}\right)} = 1.98 \times 10^{-2},\text{ so we can safely ignore relativistic effects.}$$

de Broglie wavelength:

$$\lambda = \frac{h}{p} = \frac{h}{m_e v} = \frac{6.63 \times 10^{-34}\,\text{J}\cdot\text{s}}{(9.11 \times 10^{-31}\,\text{kg})\left(5.927 \times 10^6 \frac{\text{m}}{\text{s}}\right)} = \boxed{1.23 \times 10^{-10}\,\text{m}}$$

Part d)

Speed:

$$K = \frac{1}{2}m_e v^2$$

$$v = \sqrt{\frac{2K}{m_e}} = \sqrt{\frac{2(1.60 \times 10^{-16}\,\text{J})}{9.11 \times 10^{-31}\,\text{kg}}} = 1.874 \times 10^7 \frac{\text{m}}{\text{s}}$$

Ratio of v to c:

$$\frac{v}{c} = \frac{\left(1.874 \times 10^7 \frac{\text{m}}{\text{s}}\right)}{\left(3.00 \times 10^8 \frac{\text{m}}{\text{s}}\right)} = 6.25 \times 10^{-2}$$

This speed is around 6% of the speed of light. For higher kinetic energies, we should start including relativistic effects in our calculations.

de Broglie wavelength:

$$\lambda = \frac{h}{p} = \frac{h}{m_e v} = \frac{6.63 \times 10^{-34}\,\text{J}\cdot\text{s}}{(9.11 \times 10^{-31}\,\text{kg})\left(1.874 \times 10^7 \frac{\text{m}}{\text{s}}\right)} = \boxed{3.88 \times 10^{-11}\,\text{m}}$$

Part e)

Speed from the relativistic kinetic energy:

$$K = (\gamma - 1)m_e c^2 = \left(\frac{1}{\sqrt{1 - \left(\frac{v}{c}\right)^2}} - 1\right) m_e c^2$$

$$K + m_e c^2 = \frac{m_e c^2}{\sqrt{1 - \left(\frac{v}{c}\right)^2}}$$

$$\sqrt{1 - \left(\frac{v}{c}\right)^2} = \frac{m_e c^2}{K + m_e c^2}$$

$$1 - \left(\frac{v}{c}\right)^2 = \left(\frac{m_e c^2}{K + m_e c^2}\right)^2$$

$$v = c\sqrt{1 - \left(\frac{m_e c^2}{K + m_e c^2}\right)^2}$$

$$= \left(3.00 \times 10^8 \frac{\text{m}}{\text{s}}\right)\sqrt{1 - \left(\frac{(9.11 \times 10^{-31}\text{ kg})\left(3.00 \times 10^8 \frac{\text{m}}{\text{s}}\right)^2}{(1.60 \times 10^{-13}\text{ J}) + (9.11 \times 10^{-31}\text{ kg})\left(3.00 \times 10^8 \frac{\text{m}}{\text{s}}\right)^2}\right)^2}$$

$$= 2.82256 \times 10^8 \frac{\text{m}}{\text{s}}$$

Ratio of v to c:

$$\frac{v}{c} = \frac{\left(2.82256 \times 10^8 \frac{\text{m}}{\text{s}}\right)}{\left(3.00 \times 10^8 \frac{\text{m}}{\text{s}}\right)} = 0.94085$$

Relativistic gamma:

$$\gamma = \frac{1}{\sqrt{1 - \left(\frac{v}{c}\right)^2}} = \frac{1}{\sqrt{1 - (0.94085)^2}} = 2.951$$

de Broglie wavelength:

$$\lambda = \frac{h}{p} = \frac{h}{\gamma m_e v} = \frac{6.63 \times 10^{-34}\text{ J} \cdot \text{s}}{(2.951)(9.11 \times 10^{-31}\text{ kg})\left(2.82256 \times 10^8 \frac{\text{m}}{\text{s}}\right)} = \boxed{8.74 \times 10^{-13}\text{ m}}$$

Part f)

Speed:

$$v = c\sqrt{1 - \left(\frac{m_e c^2}{K + m_e c^2}\right)^2}$$

$$= \left(3.00 \times 10^8 \frac{\text{m}}{\text{s}}\right)\sqrt{1 - \left(\frac{(9.11 \times 10^{-31}\text{ kg})\left(3.00 \times 10^8 \frac{\text{m}}{\text{s}}\right)^2}{(1.60 \times 10^{-10}\text{ J}) + (9.11 \times 10^{-31}\text{ kg})\left(3.00 \times 10^8 \frac{\text{m}}{\text{s}}\right)^2}\right)^2}$$

$$= 2.9999996 \times 10^8 \frac{\text{m}}{\text{s}}$$

Ratio of v to c:

$$\frac{v}{c} = \frac{\left(2.9999996 \times 10^8 \frac{m}{s}\right)}{\left(3.00 \times 10^8 \frac{m}{s}\right)} = 0.99999987$$

Relativistic gamma:

$$\gamma = \frac{1}{\sqrt{1 - \left(\frac{v}{c}\right)^2}} = \frac{1}{\sqrt{1 - (0.99999987)^2}} = 1961$$

de Broglie wavelength:

$$\lambda = \frac{h}{p} = \frac{h}{\gamma m_e v} = \frac{6.63 \times 10^{-34} \text{ J} \cdot \text{s}}{(1961)(9.11 \times 10^{-31} \text{ kg})\left(2.9999996 \times 10^8 \frac{m}{s}\right)} = \boxed{1.24 \times 10^{-14} \text{ m}}$$

REFLECT

The de Broglie wavelength is inversely proportional to the speed of the particle; the kinetic energy is proportional to the speed of the particle. Therefore, as the kinetic energy increases, the de Broglie wavelength of the particle decreases.

Get Help: P'Cast 26.3 – Finding the Wavelength of a Room-Temperature Neutron

26.73

SET UP

The wavelength emitted by an electron flipping spin states in the ground state of hydrogen is $\lambda = 0.21$ m. The energy difference between energy states is equal to the energy of the photon that is emitted, $E_{\text{photon}} = \frac{hc}{\lambda}$.

SOLVE

$$\Delta E = E_{\text{photon}} = \frac{hc}{\lambda} = \frac{(6.63 \times 10^{-34} \text{ J} \cdot \text{s})\left(3.00 \times 10^8 \frac{m}{s}\right)}{0.21 \text{ m}}$$
$$= \boxed{9.5 \times 10^{-21} \text{ J}}$$

REFLECT

Light with a wavelength of 21 cm is in the microwave region of the spectrum.

Get Help: P'Cast 26.4 – Photon Possibilities

26.77

SET UP

The Balmer series results from transitions of electrons in hydrogen from higher energy states to the $n = 2$ level. The Rydberg formula, $\frac{1}{\lambda} = R_H\left(\frac{1}{n^2} - \frac{1}{m^2}\right)$, where $R_H = 1.097 \times 10^7 \text{ m}^{-1}$, gives the wavelengths of the photons associated with a transition from the mth to the nth energy level.

SOLVE
Rydberg formula:

$$\frac{1}{\lambda} = R_H\left(\frac{1}{n^2} - \frac{1}{m^2}\right)$$

From $m = 3$ to $n = 2$:

$$\lambda = \left[R_H\left(\frac{1}{n^2} - \frac{1}{m^2}\right)\right]^{-1} = \left[(1.097 \times 10^7 \text{ m}^{-1})\left(\frac{1}{(2)^2} - \frac{1}{(3)^2}\right)\right]^{-1}$$

$$= \boxed{6.563 \times 10^{-7} \text{ m} = 656.3 \text{ nm}}$$

From $m = 4$ to $n = 2$:

$$\lambda = \left[R_H\left(\frac{1}{n^2} - \frac{1}{m^2}\right)\right]^{-1} = \left[(1.097 \times 10^7 \text{ m}^{-1})\left(\frac{1}{(2)^2} - \frac{1}{(4)^2}\right)\right]^{-1}$$

$$= \boxed{4.862 \times 10^{-7} \text{ m} = 486.2 \text{ nm}}$$

From $m = 5$ to $n = 2$:

$$\lambda = \left[R_H\left(\frac{1}{n^2} - \frac{1}{m^2}\right)\right]^{-1} = \left[(1.097 \times 10^7 \text{ m}^{-1})\left(\frac{1}{(2)^2} - \frac{1}{(5)^2}\right)\right]^{-1}$$

$$= \boxed{4.341 \times 10^{-7} \text{ m} = 434.1 \text{ nm}}$$

From $m = 6$ to $n = 2$:

$$\lambda = \left[R_H\left(\frac{1}{n^2} - \frac{1}{m^2}\right)\right]^{-1} = \left[(1.097 \times 10^7 \text{ m}^{-1})\left(\frac{1}{(2)^2} - \frac{1}{(6)^2}\right)\right]^{-1}$$

$$= \boxed{4.102 \times 10^{-7} \text{ m} = 410.2 \text{ nm}}$$

REFLECT
The first four lines in the Balmer series are in the visible range, ranging from red to blue to violet; the remaining ones are in the ultraviolet region.

Get Help: P'Cast 26.4 – Photon Possibilities

26.81

SET UP
The standard form of the Rydberg formula gives the wavelength of the emitted photons. Using $c = \lambda f$ we can express the Rydberg formula in terms of the frequency of the emitted photons.

SOLVE

$$\frac{1}{\lambda} = R_H\left(\frac{1}{n^2} - \frac{1}{m^2}\right)$$

$$f = \frac{c}{\lambda} = cR_H\left(\frac{1}{n^2} - \frac{1}{m^2}\right) = \left(2.9979 \times 10^8 \frac{\text{m}}{\text{s}}\right)(1.09737 \times 10^7 \text{ m}^{-1})\left(\frac{1}{n^2} - \frac{1}{m^2}\right)$$

$$\boxed{f = (3.2898 \times 10^{15} \text{ Hz})\left(\frac{1}{n^2} - \frac{1}{m^2}\right)}$$

REFLECT

Comparing the frequency of photons is a bit more intuitive than comparing their wavelengths since the energy is directly proportional to the frequency, but inversely proportional to the wavelength.

Get Help: P'Cast 26.4 – Photon Possibilities

26.85

SET UP

The speed of an electron in the nth Bohr orbit is given by $v_n = \dfrac{Z\hbar}{na_0 m_e}$ where $a_0 = 0.529 \times 10^{-10}$ m. Since we are examining a hydrogen atom, $Z = 1$. The angular momentum of an electron in the nth Bohr orbit is $L_n = n\hbar$, where $\hbar = \dfrac{h}{2\pi}$. To find these values for an electron in the tenth Bohr orbit, we should plug in $n = 10$ and solve.

SOLVE

Speed of an electron in the tenth Bohr orbit of hydrogen:

$$v_n = \frac{Z\hbar}{na_0 m_e} = \frac{Zh}{2\pi n a_0 m_e}$$

$$v_{10} = \frac{(1)h}{2\pi(10)a_0 m_e} = \frac{6.63 \times 10^{-34} \text{ J}\cdot\text{s}}{20\pi(0.529 \times 10^{-10} \text{ m})(9.11 \times 10^{-31} \text{ kg})}$$

$$= \boxed{2.19 \times 10^5 \frac{\text{m}}{\text{s}}}$$

Angular momentum of an electron in the tenth Bohr orbit:

$$L_n = n\hbar = \frac{nh}{2\pi}$$

$$L_{10} = \frac{(10)h}{2\pi} = \frac{5h}{\pi} = \frac{5(6.63 \times 10^{-34} \text{ J}\cdot\text{s})}{\pi} = \boxed{1.06 \times 10^{-33} \text{ J}\cdot\text{s}}$$

REFLECT

The general expression for the quantized radii of electron orbits is $r_n = \dfrac{n^2 a_0}{Z}$, where Z is the atomic number; for hydrogen, $Z = 1$.

Get Help: Interactive Example – Quantum Mechanical Atom
P'Cast 26.5 – Lowest Energy Level of a Lithium Ion

26.89

SET UP

A hydrogen atom has an electron in the $n = 2$ state when it absorbs a photon, which promotes the electron to the $n = 4$ state. This electron will then emit light and relax back down to the $n = 1$ state. We can use the Rydberg equation to calculate the wavelength of the absorbed photon, as well as the wavelengths of the possible emitted photons. Since the

Rydberg equation is used to calculate the wavelength of the *emitted* photon (*i.e.*, $n > m$), we need to either take the absolute value or reverse the terms to correctly calculate the wavelength of the *absorbed* photon.

SOLVE

Part a)

$$\frac{1}{\lambda} = R_H\left(\frac{1}{n^2} - \frac{1}{m^2}\right)$$

$$\lambda = \left[R_H\left(\frac{1}{n^2} - \frac{1}{m^2}\right)\right]^{-1} = \left|\left[(1.097 \times 10^7 \text{ m}^{-1})\left(\frac{1}{4^2} - \frac{1}{2^2}\right)\right]^{-1}\right|$$

$$= \boxed{4.862 \times 10^{-7} \text{ m} = 486.2 \text{ nm}}$$

Part b)

Emission process #1: $n = 4 \to n = 3 \to n = 2 \to n = 1$

$$\lambda_{4 \to 3} = \left[(1.097 \times 10^7 \text{ m}^{-1})\left(\frac{1}{3^2} - \frac{1}{4^2}\right)\right]^{-1} = 1.875 \times 10^{-6} \text{ m} = \boxed{1875 \text{ nm}}$$

$$\lambda_{3 \to 2} = \left[(1.097 \times 10^7 \text{ m}^{-1})\left(\frac{1}{2^2} - \frac{1}{3^2}\right)\right]^{-1} = 6.563 \times 10^{-7} \text{ m} = \boxed{656.3 \text{ nm}}$$

$$\lambda_{2 \to 1} = \left[(1.097 \times 10^7 \text{ m}^{-1})\left(\frac{1}{1^2} - \frac{1}{2^2}\right)\right]^{-1} = 1.215 \times 10^{-7} \text{ m} = \boxed{121.5 \text{ nm}}$$

Emission process #2: $n = 4 \to n = 2 \to n = 1$

$$\lambda_{4 \to 2} = \left[(1.097 \times 10^7 \text{ m}^{-1})\left(\frac{1}{2^2} - \frac{1}{4^2}\right)\right]^{-1} = 4.862 \times 10^{-6} \text{ m} = \boxed{4862 \text{ nm}}$$

$$\lambda_{2 \to 1} = \boxed{121.5 \text{ nm}}$$

Emission process #3: $n = 4 \to n = 3 \to n = 1$

$$\lambda_{4 \to 3} = \boxed{1875 \text{ nm}}$$

$$\lambda_{3 \to 1} = \left[(1.097 \times 10^7 \text{ m}^{-1})\left(\frac{1}{1^2} - \frac{1}{3^2}\right)\right]^{-1} = 1.026 \times 10^{-7} \text{ m} = \boxed{102.6 \text{ nm}}$$

Emission process #4: $n = 4 \to n = 1$

$$\lambda_{4 \to 2} = \left[(1.097 \times 10^7 \text{ m}^{-1})\left(\frac{1}{1^2} - \frac{1}{4^2}\right)\right]^{-1} = 9.723 \times 10^{-8} \text{ m} = \boxed{97.23 \text{ nm}}$$

REFLECT

We don't know ahead of time which relaxation process the electron will undergo on its way back to the $n = 1$ state; all are possible, but some are more probable than others.

Get Help: P'Cast 26.5 – Lowest Energy Level of a Lithium Ion

Chapter 27
Nuclear Physics

Conceptual Questions

27.7 The central ideas of binding energy of nuclear reactions would not have been possible to develop without the $E_0 = mc^2$ concept. The Manhattan project in World War II, where nuclear weapons were designed and successfully tested, would not have been possible had it not been for Einstein's theory of the equivalence of mass and energy.

Get Help: P'Cast 27.3 – The Binding Energy of ^{4}He

27.11 Consider the typical beta decay of a neutron: n → p + e⁻ + $\bar{\nu}_e$. If the electron (e⁻) were the only decay product, application of conservation of energy and momentum to the two-body decay would require that the e⁻ particle be ejected with a single unique energy. Instead, we observe experimentally that e⁻ particles are produced with energies that range from zero to a maximum value. Further, because the original neutron had spin 1/2, conservation of angular momentum would be violated if the final decay products consisted of only the two particles p and e⁻, each with spin 1/2.

Get Help: P'Cast 27.5 – Technetium
P'Cast 27.6 – Alpha Decay of ^{238}U
P'Cast 27.7 – Ötzi the Iceman

27.15 Atomic masses are used instead of nuclear masses because the masses of neutral atoms have been well measured, but precise measurements of atomic nuclei are difficult to obtain. Because the atomic masses include Z electron masses in their values, the mass of a beta particle is already taken into account. There is one electron mass included in the atomic mass of the proton. There is no need to add in another one to account for the beta particle.

Get Help: P'Cast 27.5 – Technetium
P'Cast 27.6 – Alpha Decay of ^{238}U
P'Cast 27.7 – Ötzi the Iceman

Multiple-Choice Questions

27.19 D (Both fusion and fission release energy). Energy is released in either process in order to form more stable products.

27.23 B (less than). Energy is released in a spontaneous fusion process. Mass and energy are equivalent through $E_0 = mc^2$.

Estimation/Numerical Questions

27.29 According to Example 27-2, the density of all atomic nuclei is 2.3×10^{17} kg/m³. The average radius of an atom is 10^{-10} m. The atomic mass of carbon is 2×10^{-26} kg.

$$\rho = \frac{m}{V} = \frac{m}{\frac{4}{3}\pi r^3} = \frac{3m}{4\pi r^3} = \frac{3(2 \times 10^{-26}\text{kg})}{4\pi(10^{-10}\text{ m})^3} = 4.8 \times 10^3 \frac{\text{kg}}{\text{m}^3}$$

Ratio of density of atomic nucleus to atom:

$$\frac{2.3 \times 10^{17} \frac{\text{kg}}{\text{m}^3}}{4.8 \times 10^3 \frac{\text{kg}}{\text{m}^3}} \approx 5 \times 10^{13}$$

Thus, an atomic nucleus is approximately 5×10^{13} times more dense than an atom.

Get Help: P'Cast 27.1 – Nuclear Radii
P'Cast 27.2 – Nuclear Density

27.35

$$\frac{N(t)}{N_0} = \left[\frac{1}{2}\right]^{\frac{t}{\tau_{1/2}}}$$

$$t = \tau_{1/2} \frac{\ln\left(\frac{N(t)}{N_0}\right)}{\ln\left(\frac{1}{2}\right)}$$

Part a)
$$t = (1 \text{ d}) \frac{\ln(0.625)}{\ln\left(\frac{1}{2}\right)} = 0.678 \text{ d}$$

Part b)
$$t = (1 \text{ d}) \frac{\ln(0.0625)}{\ln\left(\frac{1}{2}\right)} = 4.00 \text{ d}$$

Get Help: P'Cast 27.5 – Technetium
P'Cast 27.6 – Alpha Decay of ^{238}U
P'Cast 27.7 – Ötzi the Iceman

Problems

27.41

SET UP

A nucleus is approximately spherical with a radius $r = r_0 A^{\frac{1}{3}}$, where $r_0 = 1.2 \times 10^{-15}$ m and A is the mass number. The mass density of the nucleus is equal to its mass divided by its volume. We can approximate the total mass of the nucleus as A multiplied by the average mass of a nucleon. The mass of a proton is $m_p = 1.6726 \times 10^{-27}$ kg, and the mass of a neutron is $m_n = 1.6749 \times 10^{-27}$ kg.

SOLVE

$$\rho = \frac{m}{V} = \frac{Am_{avg}}{\left(\frac{4}{3}\pi r^3\right)} = \frac{3A\left(\frac{m_p + m_n}{2}\right)}{4\pi(r_0 A^{\frac{1}{3}})^3} = \frac{3(m_p + m_n)}{8\pi r_0^3}$$

$$= \frac{3((1.6726 \times 10^{-27} \text{ kg}) + (1.6749 \times 10^{-27} \text{ kg}))}{8\pi(1.2 \times 10^{-15} \text{ m})^3} = \boxed{2.3 \times 10^{17} \frac{\text{kg}}{\text{m}^3}}$$

$$2.3 \times 10^{17} \frac{\text{kg}}{\text{m}^3} \times \frac{2.2 \text{ lb}}{1 \text{ kg}} \times \frac{1 \text{ ton}}{2000 \text{ lb}} \times \left(\frac{1 \text{ m}}{100 \text{ cm}}\right)^3 \times \left(\frac{2.54 \text{ cm}}{1 \text{ in}}\right)^3 = \boxed{4.2 \times 10^9 \frac{\text{tons}}{\text{in}^3}}$$

REFLECT

The nucleus is extremely dense, so an answer on the order of 10^{17} is reasonable. Note that the density did not depend upon the mass number.

Get Help: P'Cast 27.1 – Nuclear Radii
P'Cast 27.2 – Nuclear Density

27.43

SET UP

Carbon-12 has a mass of $m_{atom} = 12.000000$ u and is made up of 6 neutrons, 6 protons, and 6 electrons. The binding energy of the atom is the difference in energy between the component parts of the atom and the atom itself, $E_B = (Nm_n + Zm_{^1H} - m_{atom})c^2$, where $m_n = 1.008665$ u and $m_{^1H} = 1.007825$ u. The conversion between u and MeV is 1 u = 931.494 MeV/c^2.

SOLVE

$$E_B = (Nm_n + Zm_{^1H} - m_{atom})c^2$$

$$E_B = (6(1.008665 \text{ u}) + 6(1.007825 \text{ u}) - (12.000000 \text{ u}))c^2$$

$$= (0.098940 \text{ u})c^2 \times \frac{\left(931.494 \frac{\text{MeV}}{c^2}\right)}{1 \text{ u}} = \boxed{92.162 \text{ MeV}}$$

REFLECT

The larger the binding energy is, the more stable the nucleus.

Get Help: P'Cast 27.3 – The Binding Energy of ^{4}He

27.49

SET UP

Thorium-232 undergoes the following nuclear fission reaction: ^{232}Th + n → ^{99}Kr + ^{124}Xe + __. We can determine the missing product by calculating its expected atomic number Z and mass number A. The sum of the atomic numbers of the products must equal the sum of the atomic numbers of the reactants; the same must be true for the mass numbers. The atomic numbers of thorium, krypton, and xenon are 90, 36, and 54, respectively. The mass numbers of each species are to the top left of the atomic symbol; a neutron has a mass number of 1. The energy released by the reaction is equal to the total energy of the reactants minus the total energy of the products. The atomic masses are thorium-232 = 232.038051 u, a neutron = 1.008665 u, krypton-99 = 98.957606 u, and xenon-124 = 123.905894 u.

SOLVE

Atomic number:

$$Z = 90 - (36 + 54) = 0$$

Since $Z = 0$, the missing product will be a neutron.

Mass number:

$$A = 232 + 1 - (99 + 124) = 10$$

The total mass number of the missing product must be 10. We already know it is a neutron, which has a mass number of 1, so the missing product must be 10 neutrons, or $\boxed{10n}$.

(energy released) = (total binding energy of fragments) − (binding energy of original ^{232}Th nucleus)

Binding energies of individual nuclei:

$$E_B = (Nm_n + Zm_{^1H} - m_{atom})c^2$$

$$E_{B,^{232}Th} = N_{^{232}Th}m_n + Z_{^{232}Th}m_{^1H} - m_{^{232}Th}$$

$$E_{B,^{124}Xe} = N_{^{124}Xe}m_n + Z_{^{124}Xe}m_{^1H} - m_{^{124}Xe}$$

$$E_{B,^{99}Kr} = N_{^{99}Kr}m_n + Z_{^{99}Kr}m_{^1H} - m_{^{99}Kr}$$

Energy released:

$$E_{released} = E_{B,^{99}Kr} + E_{B,^{124}Xe} - E_{B,^{232}Th}$$

$$E_{released} = [(N_{^{99}Kr} + N_{^{124}Xe} - N_{^{232}Th})m_n + (Z_{^{99}Kr} + Z_{^{124}Xe} - Z_{^{232}Th})m_{^1H} - (m_{^{99}Kr} + m_{^{124}Xe} - m_{^{232}Th})]c^2$$

$$E_{released} = [(63 + 70 - 142)m_n + (36 + 54 - 90)m_{^1H} - (m_{^{99}Kr} + m_{^{124}Xe} - m_{^{232}Th})]c^2$$

$$E_{released} = [(-9)(1.008665 \text{ u}) + (0)(1.007825 \text{ u}) - (98.957606 \text{ u} + 123.905894 \text{ u} - 232.038051)]c^2$$

$$E_{released} = (0.096566 \text{ u})c^2 \times \frac{\left(931.494 \frac{\text{MeV}}{c^2}\right)}{1 \text{ u}}$$

$$E_{released} = \boxed{89.951 \text{ MeV}}$$

REFLECT

The total number of protons and the total number of neutrons must remain constant.

Get Help: P'Cast 27.4 − Spontaneous Uranium Fission

27.55

SET UP

In Section 27-4, it is stated that the fission of uranium-235 releases 185×10^6 eV of energy. In order to determine how many kilograms of uranium-235 are necessary to produce 1000×10^6 W of power continuously for one year, we must first determine the total amount of energy required by multiplying the power by the time interval. Each fission reaction uses 1 nucleus; we can use Avogadro's number and the molar mass of uranium-235 in order to calculate the mass.

SOLVE
Total energy required:

$$P = \frac{E}{\Delta t}$$

$$E = P\Delta t = (1000 \times 10^6 \text{ W})\left(1 \text{ y} \times \frac{365.25 \text{ d}}{1 \text{ y}} \times \frac{24 \text{ h}}{1 \text{ d}} \times \frac{3600 \text{ s}}{1 \text{ h}}\right) = 3.1558 \times 10^{16} \text{ J}$$

Total number of fission reactions necessary:

$$\frac{3.1558 \times 10^{16} \text{ J}}{185 \times 10^6 \text{ eV}} \times \frac{1 \text{ eV}}{1.602 \times 10^{-19} \text{ J}} = 1.065 \times 10^{27} \text{ reactions}$$

Mass of uranium-235 required:

$$1.065 \times 10^{27} \text{ reactions} \times \frac{1 \text{ nucleus U-235}}{1 \text{ reaction}} \times \frac{235 \text{ g}}{6.02 \times 10^{23} \text{ nuclei}} = \boxed{4.16 \times 10^5 \text{ g} \approx 4 \times 10^2 \text{ kg}}$$

REFLECT

We've assumed that the fission reaction is 100% efficient, which means our answer is the *minimum* mass required.

Get Help: P'Cast 27.4 – Spontaneous Uranium Fission

27.61

SET UP

The deuterium–tritium (D–T) fusion reaction forms helium-4 and a neutron from deuterium and tritium: D + T → ^{4}He + n; the reaction releases about 20×10^6 eV of energy. We can use this information, along with Avogadro's number and the molar mass of tritium (1 mol = 3.016 g), in order to calculate the amount of tritium necessary to create 10^{14} J of energy.

SOLVE

$$10^{14} \text{ J} \times \frac{1 \text{ eV}}{1.602 \times 10^{-19} \text{ J}} \times \frac{1 \text{ nucleus T}}{20 \times 10^6 \text{ eV}} \times \frac{3.016 \text{ g}}{6.02 \times 10^{23} \text{ nuclei T}}$$

$$= \boxed{156 \text{ g} \approx 200 \text{ g (rounded to one significant figure)}}$$

REFLECT

This reaction is the simplest fusion reaction to perform on Earth and, because of this, has great promise for potential future sources of nuclear energy.

27.65

SET UP

A certain radioisotope has a decay constant of $\lambda = 0.00334 \text{ s}^{-1}$. The half-life is equal to $\tau_{1/2} = \dfrac{\ln 2}{\lambda}$.

SOLVE

$$\tau_{1/2} = \frac{\ln 2}{\lambda} = \frac{\ln 2}{0.00334 \text{ s}^{-1}} = \boxed{208 \text{ s}} \times \frac{1 \text{ h}}{3600 \text{ s}} \times \frac{1 \text{ d}}{24 \text{ h}} = \boxed{0.00240 \text{ d}}$$

REFLECT

This is about 3.5 min.

Get Help: P'Cast 27.5 – Technetium
P'Cast 27.6 – Alpha Decay of ^{238}U
P'Cast 27.7 – Ötzi the Iceman

27.69

SET UP

A patient is injected with 7.88 μCi of iodine-131 that has a half-life of $\tau_{1/2} = 8.02$ days. We want to calculate the expected decay rate in the patient's thyroid after 30 days. The decay rate exponentially decays with time, $R = R_0 e^{-\lambda t}$, where R_0 is the initial rate and $\lambda = \dfrac{\ln 2}{\tau_{1/2}}$. Only 90% of the initial dose makes its way to the thyroid, which means $R_0 = (0.90)(7.88 \, \mu\text{Ci})$.

SOLVE

Decay constant:

$$\lambda = \frac{\ln 2}{\tau_{1/2}} = \frac{\ln 2}{8.02 \text{ d}} = 0.0864 \text{ d}^{-1}$$

Decay rate after 30 days:

$$R = R_0 e^{-\lambda t} = (0.90)(7.88 \, \mu\text{Ci}) e^{-(0.0864 \text{ d}^{-1})(30 \text{ d})} = \boxed{0.5 \, \mu\text{Ci}}$$

REFLECT

The number of decays per second depends upon the number of atoms present; if the number of atoms decreases exponentially with time, so too should the decay rate.

Get Help: P'Cast 27.5 – Technetium
P'Cast 27.6 – Alpha Decay of ^{238}U
P'Cast 27.7 – Ötzi the Iceman

27.73

SET UP

A sample of radon-222 has a decay rate of $R = 485$ counts/min. The decay rate of the sample is equal to the product of the decay constant λ and the number of nuclei N. The half-life of radon-222 from Appendix C is $\tau_{1/2} = 3.823$ days. After finding λ from the half-life, we can divide the decay rate by the decay constant in order to calculate the number of nuclei in the sample.

SOLVE
Decay constant:

$$\lambda = \frac{\ln 2}{\tau_{1/2}} = \frac{\ln 2}{\left(3.823 \text{ d} \times \frac{24 \text{ h}}{1 \text{ d}} \times \frac{3600 \text{ s}}{1 \text{ h}}\right)} = 2.0985 \times 10^{-6} \text{ s}^{-1}$$

Number of nuclei:

$$R = \lambda N$$

$$N = \frac{R}{\lambda} = \frac{\left(485 \frac{\text{counts}}{\text{min}} \times \frac{1 \text{ min}}{60 \text{ s}}\right)}{2.0985 \times 10^{-6} \text{ s}^{-1}} = \boxed{3.85 \times 10^6 \text{ nuclei}}$$

REFLECT
The units of "counts/second" are equivalent to becquerels, as long as the detector detects 100% of the radioactive decays.

Get Help: P'Cast 27.5 – Technetium
P'Cast 27.6 – Alpha Decay of ^{238}U
P'Cast 27.7 – Ötzi the Iceman

27.79

SET UP

The approximate radius of a uranium-238 nucleus is given by $r = r_0 A^{\frac{1}{3}}$, where $r_0 = 1.2$ fm and A is the mass number. The magnitude of the electric force between two protons located at opposite ends of the nucleus can be calculated from Coulomb's law. If this were the only force acting on the protons, we can calculate their resulting acceleration from Newton's second law. Finally, the nucleus is held together by the strong nuclear force, which allows the nucleons to fuse together to form nuclei.

SOLVE
Part a)

$$r = r_0 A^{\frac{1}{3}} = (1.2 \text{ fm})(238)^{\frac{1}{3}} = \boxed{7.44 \text{ fm} \approx 7.4 \text{ fm}}$$

Part b)

$$F_{\text{electric}} = \frac{k|q_1||q_2|}{r^2}$$

$$F_{\text{electric}} = \frac{k(e)(e)}{r^2} = \frac{\left(8.99 \times 10^9 \frac{\text{N} \cdot \text{m}^2}{\text{C}^2}\right)(1.60 \times 10^{-19} \text{ C})^2}{(2(7.44 \times 10^{-15} \text{ m}))^2} = \boxed{1.04 \text{ N} \approx 1.0 \text{ N}}$$

Part c)

$$\sum F_{\text{ext}} = F_{\text{electric}} = m_p a$$

$$a = \frac{F_{\text{electric}}}{m_p} = \frac{1.04 \text{ N}}{(1.67 \times 10^{-27} \text{ kg})} = \boxed{6.2 \times 10^{26} \frac{\text{m}}{\text{s}^2}}$$

124 Chapter 27 Nuclear Physics

Part d) The strong nuclear force holds them together.

REFLECT
The weight of a proton is about 10^{-26} N, which means the electrostatic force between the two farthest protons in the nucleus is 26 orders of magnitude larger! For a comparison, the mass of Earth is "only" 23 orders of magnitude larger than the mass of a person.

Get Help: P'Cast 27.1 – Nuclear Radii
P'Cast 27.2 – Nuclear Density

27.83

SET UP
An isotope of element 117 was created that had 176 neutrons and a half-life of $\tau_{1/2} = 14 \times 10^{-3}$ s. The mass number A of the isotope is the sum of the atomic number and the number of neutrons. The approximate radius of the nucleus is given by $r = r_0 A^{\frac{1}{3}}$, where $r_0 = 1.2$ fm. The fraction of the sample that remains as a function of time is related to the half-life by $N(t) = N_0 e^{-\lambda t}$, where $\tau_{1/2} = \dfrac{\ln 2}{\lambda}$; we can use this expression to calculate the percent of the created isotope left after 1.0 s.

SOLVE

Part a)

Mass number:

$$A = N + Z = 176 + 117 = 293$$

Radius of the nucleus:

$$r = r_0 A^{\frac{1}{3}} = (1.2 \text{ fm})(293)^{\frac{1}{3}} = \boxed{8.0 \text{ fm}}$$

Part b)

$$N(t) = N_0 e^{-\lambda t}$$

$$N(t) = N_0 e^{-\left(\frac{\ln 2}{\tau_{1/2}}\right)t}$$

$$\frac{N(t)}{N_0} = \left(\frac{1}{2}\right)^{\frac{t}{\tau_{1/2}}}$$

$$\frac{N(1.0 \text{ s})}{N_0} = \left[\frac{1}{2}\right]^{\frac{1.0 \text{ s}}{14 \times 10^{-3} \text{ s}}} = 3.1 \times 10^{-22} = \boxed{3.1 \times 10^{-20} \%}$$

REFLECT
A time interval of 1.0 s is a little over 71 half-lives of the new element, so we should expect the amount of the element created to be extremely small.

Get Help: P'Cast 27.5 – Technetium
P'Cast 27.6 – Alpha Decay of ^{238}U
P'Cast 27.7 – Ötzi the Iceman

27.87

SET UP

Radiocarbon dating was used to determine the age of a bone sample found in a cave. The results showed that the level of carbon-14 present was 2.35% of its present-day level. The fraction of the sample that remains as a function of time is related to the half-life by $N(t) = N_0 e^{-\lambda t}$, where $\tau_{1/2} = \dfrac{\ln 2}{\lambda}$. We can use this expression to calculate the age of the bones. The half-life of carbon-14 from Appendix C is $\tau_{1/2} = 5730$ y.

SOLVE

$$N(t) = N_0 e^{-\lambda t}$$

$$N(t) = N_0 e^{-\left(\frac{\ln 2}{\tau_{1/2}}\right) t}$$

$$\frac{N(t)}{N_0} = \left(\frac{1}{2}\right)^{\frac{t}{\tau_{1/2}}}$$

$$\frac{N(t)}{N_0} = \left[\frac{1}{2}\right]^{\frac{t}{\tau_{1/2}}} = 2^{-\frac{t}{\tau_{1/2}}}$$

$$\ln\left(\frac{N(t)}{N_0}\right) = -\frac{t}{\tau_{1/2}} \ln 2$$

$$t = -\tau_{1/2} \frac{\ln\left(\frac{N(t)}{N_0}\right)}{\ln 2} = -(5730 \text{ y})\frac{\ln(0.0235)}{\ln 2} = \boxed{3.10 \times 10^4 \text{ y}}$$

REFLECT

Radiocarbon dating is commonly used to date organic samples. Plants exchange carbon with the atmosphere through photosynthesis and will, therefore, contain the various isotopes of carbon in the same ratio as the atmosphere. As long as the plant is alive, the ratio should remain relatively constant. When the plant dies, photosynthesis no longer takes place, and the amount of carbon-14 will decrease with time because it is radioactive. Some animals eat plants, and other animals will eat these animals, which is how carbon-14 incorporates itself into animals.

Get Help: P'Cast 27.5 – Technetium
P'Cast 27.6 – Alpha Decay of ^{238}U
P'Cast 27.7 – Ötzi the Iceman

27.95

SET UP

We are told that electron capture by a proton is not allowed in nature, which means the reaction $e^- + p \rightarrow n + \nu_e$, where ν_e represents a particle of negligible mass called the neutrino, does not occur. We can understand why this process does not occur by calculating the energy released using the energy-mass equivalence equation $E_0 = mc^2$. Since we are interested in the energy difference, this takes the form $\Delta E = \Delta mc^2$. The masses of a neutron, proton, and electron in amu are $m_n = 1.008665$ u, $m_p = 1.007277$ u, and $m_e = 0.0005486$ u.

SOLVE

$$E_{\text{released}} = \Delta E = \Delta mc^2 = [m_e + m_p - m_n]c^2 = [(0.0005486 \text{ u}) + (1.007277 \text{ u}) - (1.008665 \text{ u})]c^2$$

$$= (-0.0008394 \text{ u})c^2 \times \frac{\left(931.494 \frac{\text{MeV}}{c^2}\right)}{1 \text{ u}} = -0.7819 \text{ MeV}$$

Since the energy released we calculated is negative, this reaction will not proceed naturally.

REFLECT

If this reaction were to occur, all low energy electrons could react with protons and turn them into neutrons. This would destroy chemistry as we know it!

Get Help: P'Cast 27.5 – Technetium
P'Cast 27.6 – Alpha Decay of ^{238}U
P'Cast 27.7 – Ötzi the Iceman

Chapter 28
Particle Physics

Conceptual Questions

28.5 There are 10 combinations: uud, udd, ddd, uuu, uus, dds, uds, uss, dss, and sss.

Multiple-Choice Questions

28.11 **B** (meson). A particle composed of a quark and an antiquark is classified as a meson.

28.15 **A** (strong). Mesons, such as neutral pions, are the exchange particles for the strong force between nucleons.

Problems

28.23

SET UP

We can determine whether or not the reaction $p \rightarrow e^+ + \gamma$ is possible by seeing if the charge, the baryon number, and the lepton number are conserved. Every matter baryon has a baryon number of $+1$, and particles that are not baryons have a baryon number of 0. A positron has an electron–lepton number of -1, and a particle that is not a lepton has an electron–lepton number of 0.

SOLVE

Charge: A proton and a positron each have a charge of $+e$. A photon is uncharged. Charge is conserved.

Baryon number: A proton has a baryon number of $+1$. A positron and a photon each have a baryon number of 0. The baryon number is not conserved.

Electron–lepton number: A proton and a photon each have an electron–lepton number of 0. A positron has an electron–lepton number of -1. The electron–lepton number is not conserved.

No, the reaction is $\boxed{\text{not possible}}$ because neither the baryon number nor the electron–lepton number is conserved.

REFLECT

The sum of the baryon numbers and lepton numbers of the particles that participate in the process must equal the sum of the values of the particles present at the end of the process.

28.29

SET UP

We can determine whether or not the reactions are possible by seeing if the charge, the baryon number, and the lepton number are conserved. Every matter baryon has a baryon number of $+1$, and particles that are not baryons have a baryon number of 0. Matter leptons have a lepton number of +1, while antimatter leptons have a lepton number of −1. Particles that are not leptons have a lepton number of 0.

SOLVE

Part a)

$$n \to p + e^- + \bar{\nu}_e$$

Charge: A neutron and an antimatter neutrino each have a charge of 0. A proton has a charge of $+e$, and an electron has a charge of $-e$. Charge is conserved.

Baryon number: A neutron and proton each have a baryon number of +1. An electron and an antimatter electron neutrino each have a baryon number of 0. The baryon number is conserved.

Electron–lepton number: A neutron and a proton each have an electron–lepton number of 0. An electron has an electron–lepton number of +1. An antimatter electron neutrino has an electron–lepton number of −1. The electron–lepton number is conserved.

This reaction is $\boxed{\text{possible}}$.

Part b)

$$\mu^- \to e^- + \bar{\nu}_e + \nu_\mu$$

Charge: A neutrino and an antimatter neutrino each have a charge of 0. A muon and an electron each have a charge of $-e$. Charge is conserved.

Baryon number: All of the particles involved have a baryon number of 0. The baryon number is conserved.

Lepton number: An electron has an electron–lepton number of +1. An antimatter electron neutrino has an electron–lepton number of −1. The electron–lepton number is conserved. A muon and a muon neutrino each have a muon–lepton number of +1. The muon–lepton number is conserved.

This reaction is $\boxed{\text{possible}}$.

Part c)

$$\pi^- \to \mu^- + \bar{\nu}_\mu$$

Charge: A negatively charged pion and a muon each have a charge of $-e$. An antimatter neutrino has a charge of 0. Charge is conserved.

Baryon number: All of the particles involved have a baryon number of 0. The baryon number is conserved.

Lepton number: A pion has a lepton number of 0. A muon has a muon–lepton number of +1. An antimatter muon neutrino has a muon–lepton number of −1. The muon–lepton number is conserved.

This reaction is $\boxed{\text{possible}}$.

REFLECT

All three reactions conserve charge, baryon number, and lepton number.

28.33

SET UP

A high-energy photon in the vicinity of a nucleus can create an electron–positron pair by pair production: $\gamma \rightarrow e^- + e^+$. The minimum energy required for this process is equal to the rest energy of an electron and a positron. The mass of an electron and the mass of a positron are both equal to $m_e = 9.11 \times 10^{-31}$ kg. Since the electron and positron are created at rest, the nucleus is necessary to conserve momentum.

SOLVE

Part a)

$$E = m_e c^2 + m_e c^2 = 2m_e c^2 = 2(9.11 \times 10^{-31} \text{ kg})\left(3.00 \times 10^8 \frac{\text{m}}{\text{s}}\right)^2$$

$$= \boxed{1.64 \times 10^{-13} \text{ J}} \times \frac{1 \text{ eV}}{1.602 \times 10^{-19} \text{ J}} = 1.02 \times 10^6 \text{ eV} = \boxed{1.02 \text{ MeV}}$$

Part b) The nucleus is required to absorb the momentum so that conservation of momentum is not violated.

REFLECT

The excess energy of a photon with an energy higher than 1.02 MeV undergoing pair production goes into the kinetic energy of the electron and the positron.

28.35

SET UP

A proton–antiproton annihilation takes place and the resulting photons have a total energy of 2.5×10^9 eV. The energy of the photons is equal to the sum of the rest energies and the kinetic energies of the proton and antiproton. The mass of a proton and the mass of an antiproton are both equal to $m_p = 1.67 \times 10^{-27}$ kg. Using energy conservation, we can calculate the kinetic energy of the proton K_p and the kinetic energy of the antiproton $K_{\bar{p}}$ if they have the same kinetic energy or if $K_p = 1.25 K_{\bar{p}}$.

SOLVE

Mass of proton in eV/c^2:

$$E = m_p c^2 = (1.67 \times 10^{-27} \text{ kg})\left(3.00 \times 10^8 \frac{\text{m}}{\text{s}^2}\right)^2$$

$$= 1.503 \times 10^{-10} \text{ J} \times \frac{1 \text{ eV}}{1.602 \times 10^{-19} \text{ J}} = 9.38 \times 10^8 \text{ eV}$$

$$m_p = 9.38 \times 10^8 \frac{\text{eV}}{c^2}$$

Part a)

$$E = m_p c^2 + m_{\bar{p}} c^2 + K_p + K_{\bar{p}} = 2m_p c^2 + 2K_p$$

$$K_p = \frac{E - 2m_p c^2}{2} = \frac{(2.5 \times 10^9 \text{ eV}) - 2\left(9.38 \times 10^8 \frac{\text{eV}}{c^2}\right)c^2}{2} = 3.12 \times 10^8 \text{ eV} \approx 0.31 \text{ GeV}$$

$$\boxed{K_p = K_{\bar{p}} = 0.31 \text{ GeV}}$$

Part b)

$$E = m_p c^2 + m_{\bar{p}} c^2 + K_p + K_{\bar{p}} = 2m_p c^2 + 1.25 K_{\bar{p}} + K_{\bar{p}}$$

$$K_{\bar{p}} = \frac{E - 2m_p c^2}{2.25} = \frac{(2.5 \times 10^9 \text{ eV}) - 2\left(9.38 \times 10^8 \frac{\text{eV}}{c^2}\right)c^2}{2.25} = 2.77 \times 10^8 \text{ eV} \approx 0.28 \text{ GeV}$$

$$\boxed{K_p = 1.25 K_{\bar{p}} = 1.25(2.77 \times 10^8 \text{ eV}) = 0.35 \text{ GeV}}$$

REFLECT

Since the proton and antiproton annihilate one another, both their kinetic energy *and* rest energy are converted into photons.

28.39

SET UP

A proton and an antiproton undergo pair annihilation and produce two photons. The Feynman diagram should have two straight lines on the left representing the proton and antiproton that meet at a vertex. Two wavy lines representing the photons should then leave the vertex headed toward the right.

SOLVE

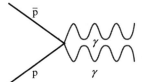

Figure 28-1 Problem 39

REFLECT

Time runs from left to right in a Feynman diagram, so the process starts with two particles and ends with two photons, as expected for an annihilation process.

28.45

SET UP

The strong force acts over distance of approximately 0.731 fm. We're interested in the nucleus that has a diameter equal to ten times this, or with a radius of $r = 3.655$ fm. We can use $r = r_0 A^{\frac{1}{3}}$, where $r_0 = 1.2$ fm, to calculate the mass number of this nucleus, and then consult a periodic table (or Appendix C) to see which element has an atomic mass similar to this. As an approximation, we can treat the outer proton of the nucleus as being repelled by all of the inner protons located at the center of the nucleus and use Coulomb's law to calculate the magnitude of this repulsion.

SOLVE

Part a)

$$r = r_0 A^{\frac{1}{3}}$$

$$A = \left(\frac{r}{r_0}\right)^3 = \left(\frac{3.655 \text{ fm}}{1.2 \text{ fm}}\right)^3 = 28.3$$

Consulting the periodic table, silicon (Si) has an atomic weight of 28.1, so a $\boxed{\text{silicon}}$ nucleus would have a diameter equal to about 10 times the range of the strong force.

Part b)

Silicon has an atomic number of 14. We will treat the outer proton as being repelled by the 13 inner protons located at the center of the nucleus:

$$F_{\text{electric}} = \frac{k|q_1||q_2|}{r^2}$$

$$F = \frac{k(e)(13e)}{r^2} = \frac{\left(8.99 \times 10^9 \frac{\text{N} \cdot \text{m}^2}{\text{C}^2}\right)(13)(1.602 \times 10^{-19} \text{ C})^2}{(3.655 \times 10^{-15} \text{ m})^2} = \boxed{225 \text{ N}}$$

REFLECT

The strong force opposes this electrical repulsion and keeps the nucleus together.